Public Landscape Integration: Public Landscape

公共景观集成
——公共景观

金盘地产传媒有限公司　策划
广州市唐艺文化传播有限公司　编著

中国林业出版社
China Forestry Publishing House

序言

随着公共景观设计在世界各城市的复苏以及人们日愈增强的意识，公共景观设计的重要性在政府的大力宣传以及支持下显得尤为突出。通过设计公共景观，城市环境得到提升，城市的身份也得以在公众面前建立，而且城市街道的形象会得到提高。注重功能性，安全性，美学功能的公共景观不仅可以促进一个城市的发展，还可以提高城市居民的生活水平。因此，公共景观设计并非只是针对一个人或是一圈人，而是适用于所有的人。

景观不仅存在公共环境中，而且建筑本身也是景观。随着人们对生活环境以及水平的要求日愈变高，人们更希望看到一些新鲜的独特的景观。本套书除了囊括国内外知名的公共景观设计，还包括各种独特的建筑景观。例如挪威 Ornesvingen 观景台，不仅拥有优美的建筑结构，而且将周围的环境映衬的更加壮丽，吸引了很多来自全世界的游客。

在城市中，街道起着决定性的作用。从当初的交通运输作用，现在的街道具有全面，强大的作用。而人们也希望穿梭于舒适，功能齐全，美观的街道。这样一来，街道设施，如：座椅，自行车停放架，候车亭，照明设施等等都有了新的角色去扮演。

公共景观设计要求公共景观和设计二者和谐统一。同时，公共空间中的设施也需要合理设计与安排。

本套书共有三册，每册书各有重点，特色。形象的图标带来更加直接，丰富的信息。第一册是"公共景观"，主要包括国内外知名的广场，公园和游乐场设计。第二册是"建筑景观"，主要包括以下三个类别：建筑造型和建筑结构，观景台，标识。创意的建筑造型和建筑结构，总能带给人新鲜的感觉。大胆的观景台设计，不仅吸引着游客，而且值得人们参考设计。极具功能作用的标识设计，形象而直接。第三册是"街道设施"，主要包括：座椅，自行车停放架，候车亭，照明设施，自动饮水器，垃圾桶和树木防护装置。这些街道设施不仅覆盖面广，而且具有强大的功能性，更加美化了城市街道。

这套书在一个更高的层次上来诠释良好公共环境的形成，可以提高人们的生活水平。特别是随着城市环境数字化的加强，舒服的公共空间设计和便利的公共设施让人们的生活变得越来越高效和充满活力。

Preface

With the revitalization of public environment design in cities around the world, as also the increased recognition of citizens, the importance of public design has emerged for the construction of a guideline which is maintained and managed by government integratively. Through the improvement for city environment by public design, the establishment of city's own identity and the creation of street image in a view of pedestrian is to get higher. A quality of a city is reflected by functionality, safety and aesthetics simultaneously. Therefore, the meaning of public environment design is not for some unique, specific or a minor group, but for several people to use or see for the creation of beautiful city landscape, which also means a design for all kind of people.

Landscape not only exists in the public environment, and the architecture itself can also be a landscape. With people's increasingly high requirement of living environment and living level, people prefer to see some fresh and unique landscape. In addition to including well-known public landscape designs, this book also includes a variety of unique architectural landscapes. For example, Norway Ornesvingen viewing platform not only has a beautiful architectural structure, but also makes the surrounding environment more spectacular, and attracts a lot of tourists from all over the world.

In a city, the street plays a decisive role. From the start, the street was just used for transportation. However, the street now has a comprehensive and powerful role. At the same time, people want to walk through a comfortable, functional and beautiful street. As a result, street furniture, such as bench, bicycle rack, bus shelter, lighting facilities, etc., have a new role to play.

The public landscape design requires public landscape and design to be of harmony and unity. At the same time, the facilities in the public space need reasonable designs and arrangements.

This book contains three volumes, each with unique focus and feature. The icon images bring more direct and rich information. Volume 1, named "Public Landscape", includes famous squares, parks and playgrounds design from all over the world. Volume 2 is "Architectural Landscape", mainly including the following three categories: Architectural Formative Arts and Structure, Observation Platform, Signage. Innovative architectural formative arts and structure always give people a fresh feeling. The bold observation platform design not only attracts tourists, but also is worth referencing. The signage is functional, image and direct. Volume 3, "Street Furniture", includes Bench, Bicycle Rack, Bus Shelters, Lighting, Drinking Fountain, Trash and Trees Protective Device. These street facilities not only cover a lot, and with powerful features, make the city streets more beautiful.

The purpose of this book is to design a book with applied examples of the amenity elements in the street furniture and to accomplish the public facilities on a higher level as one of the design method to upgrade the quality of citizens' lifestyle in the city environments. Especially with the stress of digital city environments development, comfortable public space and street furniture make citizens' lifestyle more energetic and effective.

目录
Content

第一册 公共景观
Vol.1 Public Landscape

 广场和公园
Square & Park ·· 008

梅里达工厂青年运动场
Merida Factory Youth Movement

特拉维夫港口公共空间重建
Tel Aviv Port Public Space Regeneration

台湾基隆港口
Taiwan Keelung Port

Eduard Wallnofer 广场
Eduard Wallnof Platz

Jack Evans 海港
Jack Evans Boat Harbour

Elwood 海滩
Elwood Foreshore

Pirrama 公园
Pirrama Park

蓝花楹广场
Jacaranda Square

717 Bourke 街道
717 Bourke Street

社区公园
Neighborhood Park

Banyoles 公共空间
Public Spaces in Banyoles

Rubi——市政市场大楼广场
Rubi, Municipal Market Building and Plaza

哈格里夫斯购物广场 Grounds of Hargreaves Mall	HtO 城市海滩 Popular Urban Beach HtO
Rooke 自然保护区公园 Rooke Reserve Park	糖果沙滩 Sugar Beach
Nicolai Kultur 中心 Nicolai Kultur Center	加拿大文明博物馆广场 Canadian Museum of Civilizations Plaza
Choorstraat 庭院 Choorstraat Papenhulst	达尔哥诺·玛公园 Park Diagonal Mar
Malov 轴心流畅景观 Fluent Landscapes / Malov Axis	高线公园 2 期 Section 2 of the High Line
锡姆科波浪桥 Simcoe Wavedeck	Maister 将军纪念公园 General Maister Memorial Park
马德里 RIO 景观 Madrid RIO	布里克斯顿广场 Brixton Square
城市之丘 The City Dune	铁锚公园 Anchor Park
比拉容马尔公园 Birrarung Marr Park	奥斯陆歌剧院的屋顶 Roof of Oslo Opera
五船坞广场，加菲尔德街 Five Dock Square, Garfield Street	索伯格塔及休闲区 Solberg Tower & Rest Area
墨尔本会展中心 Melbourne Convention Centre	埃斯比约海滨长廊 Esbjerg Beach Promenade
海滨公园 Waterfront Park	Ubuda 城市中心 Ubuda City Centre
巴塞罗那 El Jarddin Botanico El Jarddin Botanico de Barcelona	Dania 公园 Dania Park
蒙特伊克花园及凉亭 Montjuic Garden and Pavilion	Marina 公园 Marina Park
Ballast Point 公园 Ballast Point Park	Nou Barris 中央公园 Parc Central de Nou Barris
南森公园 Nansen Park	Arena 大道和阿姆斯特丹 Poort 商业街 Arena Boulevard / Amsterdamse Poort

罗马采石场设计
Roman Quarry Redesign

Maddern 广场
Maddern Square

M 中心
M Central

Promenade Samuel-De Champlain 长廊
Promenade Samuel-de Champlain

河滨公园凉亭
Riverside Park Pavilion

城市甲板
The City Deck

北查尔斯顿海军纪念馆
The Greater North Charleston Naval Base Memorial

东湖岸公园
The Park at Lakeshore East

菲尼克斯市政遮阴篷
Phoenix Civic Space Shade Canopies

Lillejord 休息区及人行桥
Lillejord Rest Area & Footbridge

感官花园
Sensational Garden

NPS 屋顶花园
NPS Podium Roof Garden

Perruquet 公园
Pinar del Perruquet Park

伊利街广场
Erie Street Plaza

美国迈阿密南岬公园
South Pointe Park

Freres-Charon 广场
Square des Freres-Charon

Ricard Vines 广场
Ricard Vines Square

爱沙尼亚路博物馆的露天展厅
Open Air Exhibition Grounds

海景廊架
On the Way to the Sea

9 女孩纪念馆
M9 Memorial

Holding Pattern 公共装置
Holding Pattern, PSI 2011 Installation

Dzintari 森林公园
Dzintari Forest Park

夏洛特花园
Charlotte Garden

Arboretum Klinikum 花园
Arboretum Klinikum

Gleisdreieck 公园
Park am Gleisdreieck

万桥园
Garden of 10,000 Bridges

格拉纳达论坛广场
Forum of Granada

Las Arenas 广场
Las Arenas Square

Cufar 广场
Cufar's Square

里昂草地公园
Lyon Meadows Park

公共图书馆和阅读公园
Public Library and Reading Park

Asmacati 购物中心
Asmacati Shopping Center

抱树亭
Treehugger, One Fine Day

公园中的"拼凑图"
A Patchwork in the Park

Saint Georges 广场
Saint Georges Environment

游乐场
Playground ···286

Van Beuningenplein 游乐场
Van Beuningenplein Playground

达令港广场
Darling Quarter

Zurichhorn 游乐场
The Zurichhorn Playground

PS1 跳舞的钢管
PS1 Pole Dance

火蜥蜴游乐场
Salamander Play ground

Norteland 游乐场
Norteland Playground

Rommen 学校和文化中心
Rommen School and Cultural Center

谷脊草甸游乐场
Valley Ridge Meadow Playground

Gjerdrum 高中
Gjerdrum High School

广场和公园
Square & Park

梅里达工厂青年运动场
Merida Factory Youth Movement

项目档案	Project Facts
设计：Selgas Cano 建筑事务所	Landscape Architecture: Selgas Cano Architects
面积：3 090 平方米	Site Area: 3,090 m²
完成时间：2011	Year: 2011

梅里达工厂青年运动场是一个新的城市活动中心，设计由 Selgas Cano 建筑事务所完成。

梅里达工厂青年运动场可以进行各种各样的活动，例如：溜冰、音乐会、宽带上网、艺术表演、当代舞蹈等等，并且有大量的新活动不断加入。我们不知道接下来会有什么样的新活动，但是作为建筑师，我们会在专业能力与职责范围之内，在保留原有运动的情况下让更多的新活动可以融入进来。因此，运动场被设计成一个巨大的天蓬，对整个城市开放，吸引了更多人前来休息和运动。整个运动场被一系列卵圆形的植被分隔开来，每一部分都具有专门的功能性和规则性。大面积的天蓬挡住了阳光和雨水，天蓬的表面就像是一片片云彩轻轻展开，并且以一米厚的立体网格结构构成不同的层次，透明并具有保护作用。

这个项目功能性强大。那么，准确来说，这是一个什么样的项目呢？含糊其辞来说，我们可以称之为一座建筑。但是，它所表现出来的功能性，使得我们不能将它视为简简单单的一座建筑，而应该是一个被巨大的明亮的华盖覆盖着的建筑综合体，或者说是组合有序的区块构成的一个庞大的群落。不管我们怎么称呼它，西班牙因为这个项目，而更加引人注目。建筑师们有幸邀请有名的摄影师 Iwan Baan 来拍摄整个空间。Iwan Baan 的摄影技术一流，能完美再现这个项目的风采。

Square & Park

广场和公园

Square & Park

广场和公园

Square & Park

广场和公园

Square & Park

广场和公园

Square & Park

The Merida Factory Youth Movement is a new urban center in Merida Spain designed by Selgas Cano Architects. Skatepark, broadband Internet, Contemporary Dance, Dance Funk and Hip Hop are the recognized activities composing the collective factory and will join at the consequently so called Factory of Merida. We do not know what type of filter will be applied in the future to the new forth or upcoming activities, but we intend though, within our assigned role as architects, not to have any filter to anyone. Therefore, the building is conceived as a large canopy opened to the whole city to gather anyone who may need to shelter there. This canopy is supported by a series of ovoid plant parts holding different elements of the requested functions, which are treated as independent modules able to be used separately with whole autonomy, regulated and controlled by the direction of the factory movement. The activities taking place below are covered from rain and sun by the big canopy, acting like a big termical one-meter-thick cushion, so that there will be no need to use air conditioning. The roof is understood, and extends like a light cloud, protective, translucent and constructed with a three-dimensional mesh structure one-meter-thick covering different levels.

This project accommodates a wide array of functions, everything from skateboarding to tightrope walking, but what exactly is it? It seems a bit inarticulate to call a project like the Merida Factory Youth Movement by Selgas Cano a building, because, even though it is built, it does much more than act as single building. Maybe it's better to call it a constructed body covered by a luminous canopy, or an assemblage of programmatic volumes without air conditioning. Whatever we call it, the kids in Merida, Spain are lucky to have such a great-looking place to assemble, skate and tightrope walk. The architects are also lucky to have Iwan Baan photograph the space and make it look as great as it does. It's just that his photos are always excellent.

广场和公园

特拉维夫港口公共空间重建
Tel Aviv Port Public Space Regeneration

项目档案

设计：Mayslits Kassif 建筑事务所
项目地点：以色列，特拉维夫市
面积：55 000 平方米

Project Facts

Landscape Architecture: Mayslits Kassif Architects
Location: Tel Aviv, Israel
Site Area: 55,000 m²

特拉维夫港口位于以色列最美的海滨，原先用于船舶停泊，但是在 1965 年的时候被废弃了，所以从那个时候起这个港口就被人忽视了。Mayslits Kassif 建筑事务所成功地再现了这个地方独一无二的风采，将它变为一个突出的有生气的城市地标。这个项目极具挑战性。建筑师们充分抓住这个绝佳机会，很好地处理了私人和公共环境发展的矛盾，创造了一个受欢迎的公共空间。

当地的人和其他地方的游客在这个项目完成之后都纷纷来到这里。由于受到市场的控制，城市规划受到很大的限制。但是，因为这个项目的影响力，港口周边的五公顷面积都将用于这个项目的开发。项目设计上引入了一种具有延展性的、波浪形的不分层的表面，这不仅和港口的沙丘地基相得益彰，而且为游客提供了较好的休息和活动空间。这个独一无二的城市平台，充满了主动性和艺术性，凝聚了很多人的努力与心血，具有无限的活力。

每年来到特拉维夫港口的游客差不多有 250 万，所以，这个项目在特拉维夫受到了前所未有的重视。在获得国际认可的同时，也获得了多个著名的建筑奖项，例如，获得了 2010 年 Rosa Barba 欧洲景观奖。这个港口得到了公众的喜爱，成为当地人们最喜欢的娱乐空间。作为城市的新地标，这个项目为城市海岸注入了新的生命力，并且引领了特拉维夫海岸一带一系列的公共空间项目，加速了城市和海岸的持续性发展。

Square & Park

广场和公园

Situated on one of Israels most breathtaking waterfronts, the Tel Aviv Port was plagued with neglect since 1965, when its primary use as an operational docking port was abandoned. The recently completed public space development project by Mayslits Kassif Architects managed to restore this unique part of the city, and turned it into a prominent, vivacious urban landmark. The architects viewed the project as a unique opportunity to construct a public space which challenges the common contrast between private and public development, and suggested a new agenda of hospitality for collective open spaces.

The design introduces an extensive undulating and non-hierarchical surface that acts both as a reflection of the mythological dunes on which the port was built, and as an open invitation to free interpretations and unstructured activities. Various public and social initiatives from spontaneous rallies to artistic endeavors and public acts of solidarity are now drawn to this unique urban platform, indicating the project's success in reinventing the port as a vibrant public sphere.

The design was quickly brought to life by a new management, with locals and visitors flocking to the revamped port even before the project was completed. Remarkably, despite city planning being dominated by market forces, and because of its immense popularity among the public, the project has been able to circumvent massive development schemes intended for the port's 5 hectares area.

Square & Park

023

台湾基隆港口
Taiwan Keelung Port

项目档案	Project Facts
设计：Vicente Guallart, Maria Diaz	Main Architects：Vicente Guallart, Maria Diaz
项目地点：台湾	Location：Taiwan

基隆是台北的一个港口，位于台北以北30公里处。作为亚洲最重要的港口之一，基隆无可厚非地成为了经济快速发展的载体。这个港口的功能性从另外一方面来说也限制了一个高水准的公共空间的创造。

针对这一方面，政府机构把这个项目视为建立台湾新"大门"规划中的一部分，旨在促进城市和港口的互动。在竞标阶段以及后续的建造过程中，需要解决的一个最基本的问题是怎样确定这个新的城市中心公共空间的特色，并为公众所熟知。纵观历史，亚洲城市有一个很鲜明的传统——公共空间的利用以及动态的室内室外相一致；这种传统在城市、街区、住宅或商业地块都有体现。但是，近年来经济的发展似乎将城市的发展更着重于公共空间，并且是参照美国的模式。所以，很难确认最近几年出现的城市公共空间是否与传统的动态空间相一致。而基隆港口这个项目成为了另一个进步的先例，尽管这种进步在美洲、欧洲和澳洲都已经出现。这些国家将城市与港口的互动重新定义为港口公共空间利用的最大化。这个古老的港口临近城市的中心，俨然成为娱乐、贸易、运动甚至是酒店和住宅区域。

我们的目标就是创造一个不管是在内还是外在，都具有代表性和功能性的一个城市中心综合体。在功能方面，不仅要满足基本的需求，而且要满足运动、文化、交流和其他功能，并且这些功能要和现代化的形象紧密结合。在港口的西区，各种各样的文化，行政和商业设施的连接本身是存在问题的，所以需要重新定义结构。同时也有必要增多更多的功能设施，来加强公共空间的合理利用。港口本是固定的，但是，参考类似项目的设计方案并重新确定设计想法后，一条位于城市和海岸的动态水界限形成了。在分析了各种活动的功能性之后，设计师提出了"绿廊"的设计想法。这条"绿廊"连接贸易区和车站，并且采用木质铺面，扩大了视野范围。看起来像是一种海洋植物的带有触角的叶状体，在水平面和垂直面上均折叠起来，不仅为滨海区创造了更多的休息空间，而且呈现出这个港口的名字 K-E-E-L-U-N-G 这七个字母，甚是显眼。在这个项目中，传统的东方石林被改造成依附在表面的木制折叠物，邀请游人亲身体验。

Square & Park

广场和公园

Square & Park

Keelung is the port of Taipei, the capital of the island of Taiwan. Located 30 km to the north of the capital, it is one of the most important container ports in Asia. Keelung has all the vitality of a major port, with one of the most bustling night-time markets in the Ear East and an extensive and multifarious central commercial area adjoining the port. The city nevertheless bears the traces of rapid economic growth. Its principal transportation infrastructures —— roads, railway lines, and the port itself —— continue to limit the creation of quality public spaces of in the downtown area.

In the light of this, the authorities invited projects as part of the plan to create new "Gateways" in Taiwan, oriented toward defining the interaction between the port and the city. In fact, the fundamental issue to be resolved by the various projects drawn up during the different phases of the competition and in the subsequent construction scheme was how to identify the characteristics of a new central public space for the city with which the citizens of Keelung could identify. Historically, Asian cities have a strong tradition of use of the public space and a dynamic inside-outside relationship that has generated numerous instances of cities, neighbourhoods and residential or commercial sectors of great urbanity. However, the economic development of recent years seems to have oriented urban development toward public spaces more in line with the American model, based on the habitability of air-conditioned interior spaces or urban mobility based on the car that makes the car park one of the fundamental interchanges in urban life. This makes it difficult to identify significant urban spaces created in the last few years that respond to the traditional dynamic occupation of the public space. Keelung is seeing the start of another process that is already present in most American, European and Australian cities, in which the port-city interaction is redefined in the interests of a greater public use of port spaces. In this way the historic port zones, which are normally in the proximity of central urban places, are ceded by the port to the city as a site for leisure and commercial uses, sports ports and even hotel and residential zones.

广场和公园

A city is defined internally and externally by the strength and quality of its central space. Our proposal offers the opportunity of creating a more powerful and complex urban centre that will come to constitute Keelung's symbolic space and serve to project an image of the city to the exterior.

To obtain new spaces of public amenities. A city assumes its real significance for its citizens when it has a network of amenities that satisfies not only the basic needs but also the demand for sports, culture, contact and other analogous needs. The image of modernity is closely associated with these functions, which in Keelung are still only an incipient presence. Our proposal links the access infrastructures with the zone of amenities around the Cultural Centre, creating new facilities that will serve to reinforce the centrality of the space.

Square & Park

广场和公园

In the West zone, there are a variety of cultural, administrative and commercial facilities whose integration is problematic, because of the difficulties in the way of pedestrian movement. At the same time there is a need to generate new uses and functions that will reinforce the relational contents of the public spaces, in order to provide a qualitatively richer urban framework for the citizens. This entails the restructuring of the area of amenities to the East of the port and the incorporation of new functions.

The project is based on a fixed coastline, centring the design on the creation of a dynamic line between the urban edge and the platform, reworking ideas developed in previous projects. In this case, having analysed the functioning of the various activities that come together here, the scheme proposes a pergola that provides a covered walkway extending from the commercial zone to the station, dynamically expanding this structure by way of the wooden platform. This pergola, created with a linear pattern like the tentacular fronds of a marine plant, is folded both vertically and horizontally to generate rest spaces on the seafront and to spell out the word K-E-E-L-U-N-G on the urban front. This new timber platform will thus act as an icon similar to those ferry terminals in which the name of the port is eye-catchingly displayed.

Square & Park

广场和公园

Eduard Wallnofer 广场
Eduard-Wallnofer-Platz[Landhausplatz] in Innsbruck, Tyrol

项目档案

设计：LAAC Architects zt.og, Stiefel Kramer Architecture OG(Hannes Stiefel)
项目地点：提洛尔，因斯布鲁克
面积：9 000 平方米
材料：混凝土、钢筋、玻璃
完成时间：2011

Project Facts

Landscape Architecture：LAAC Architects zt.og, Stiefel Kramer Architecture OG (Hannes Stiefel)
Location：Innsbruck, Austria Amt der Tiroler Landesregierung Abteilung Hochbau
Site Area：9,000 m²
Materials：Concrete, Steel, Glass
Year：2011

这个项目的目的在于创建一个现代的城市公共空间，为人们提供多种多样的活动，同时，需要解决各种相互矛盾的现实情况和地块的限制性。建成的广场是因斯布鲁克市面积最大的公共广场，同时因有四座纪念馆而具有象征意义。

在整修以前，这个广场感觉像是纳粹主义时期的省政府机构，而其中本来代表自由的纪念碑却呈现出法西斯意象。重新设计的目的就是纠正纪念碑所表现出来的错误概念，增强这个广场的历史重要性。整个广场的新地形为纪念馆提供了一个现代的变化的地基，并赋予纪念馆新的意义，而且方便人们进出。

新的地形实际上也是一道独特的景观。整个场地是由混凝土铺设而成的，看似一个完整的有机体，功能性

Square & Park

强大。鉴于空间的限制，功能上的要求以及形态学的考虑，广场的进入便利性和通道的规划经过了成熟的设计。这个广场为路过的人和游人提供了一个位于车站和老镇之间的公共空间。白天，广场上的光与影，还有种植的树木与广场明亮的立体静态表面形成了鲜明的对比；晚上，立体表面上的雕塑呈现出不同的影子。

在广场的北面，大楼的正前方有一个宽广的平地，这里是一个多功能的空间。一个大型的喷泉消逝了夏天的炎热。解放纪念馆南面的地形在空间上得到了充分利用。混凝土表面的质地会根据不同的几何地形区分。在很多树木下面，地面逐渐聚拢形成座位，表面被打磨光滑。

其中一座纪念碑的地基和一个新的喷泉的源头交汇在一起。水顺着阶梯流下，形成一个小水坡，孩子们可以在这里尽情地玩水嬉戏。当然，这里也有不同高度的喷泉，人们可以饮用这些水。

整个广场地面是由混凝土平板构成的，并由螺栓拉紧和巩固。每个活动区域控制在100平方米以内，散布在整个广场上。整个排水系统，包括喷泉，位于每块区域的接合处，非常隐蔽，在广场表面绝对看不到。地下车库特有的缓冲区解决了地表水的排除问题。

Goal of the project is to create a contemporary urban public space that negotiates between the various contradictory conditions and constraints of the site and establishes a stage for a new meange of urban activities characterized by a wide range of diversity. The realized project consists of a 9,000 square meters concrete floor sculpture. Eduard-Wallnofer-Platz was the largest but neglected public square in the centre of the city of Innsbruck in Tyrol, Austria. The site nevertheless kept a symbolic significance with the four memorials positioned there.

Before the transformation took place, the square's atmosphere and spatial appearance was dominated by the facing facade of the Tyrolean provincial governmental building from the period of National Socialism, and by a large scale memorial that looks like a fascist monument, which in fact and in spite of its visual appearance is a freedom monument that shall commemorate the resistance against, and the liberation from National Socialism. The intervention aims to compensate for existing misconceptions and to reinforce the monument-historical significance. The new topography of the square offers a contemporary and transformative base for the memorials and makes them accessible and regarding a new perception.

The new topography sets a landscape-like counterpart to the surrounding. But it turns into an urban sculpture through its city context, its finish in concrete and its function. Accessibility and the layout of paths result from the modulation of the surface which deals with spatial constraints, functional requirements and morphological considerations.

Pedestrians and users as well as the memorials in their role as protagonists on this new city stage allow for an operative public and open forum between main station and old town. The bright surface of the square functions as a three-dimensional projection field on which the protagonists together with the trees cause a high-contrast dynamic play of light and shadow during daytime. In front of this background the seasons are staged powerfully. Indirect light reflected from the floor sculpture directs the scenery at night times.

Square & Park

In the northern part of the square, the spacious flat area in front of the Landhaus is conceived as a generous multi-purpose event space providing the according infrastructure. A large scale fountain activates the expanded field and provides cooling-down in summertime. South of the liberation monument, the topography features a variety of spatial situations for manifold utilisations. The texture of the concrete surface varies according the type of geometrical configuration. Beneath many trees the floor continuously merges into seat accommodations with a terrazzo like polished finish.

The sculpture group of one of the monuments is integrated into the basin of a new fountain where water runs down steps cut into a slope. The shoal fountain and the water games in front of the Landhaus provide playground for children and cool down the climate in summer locally. There are drinking fountains in different heights for children and adults.

The surface of the square is realised in modulated slabs out of concrete, joined by bolts that deal with shearing forces. Infrastructural elements for the organisation of events which can take place anywhere on the square are integrated in the construction of slab——fields of max. 100 square meters. Drainage of the whole square including the fountains is located completely at the open joints between the individual fields so that there is no drainage pit visible on the whole site. An innovative buffer system allows that despite of the existence of a subterranean garage all the appearing surface water drains away within the property.

广场和公园

Jack Evans 海港
Jack Evans Boat Harbour

项目档案	Project Facts
设计：ASPECT 工作室	Landscape Architecture：ASPECT Studio
项目地点：新南威尔士，堤维德岬	Location：Tweed Heads, NSW
面积：43 000 平方米	Site Area：43,000 m²
完成时间：2011	Year：2011

Jack Evans 海港重修的一期已经完成，新的娱乐和水下游乐场已经对外开放。这个项目使得原有的公园变成一个独一无二的供市民休闲的地方。活动设施方面设计了大量的人行道，单车道和阶梯，以及供市民游泳和划船的设施，主要目的在于锻炼人的体力。在这个 43 000 平方米的海港上，也有供市民开会、周末交易、举行纪念仪式和小孩子玩耍的空间。这个海港也是著名的游览圣地，集多样性，活力，教育性和娱乐性于一体，带动了堤维德岬的经济发展。

这个项目是征集了社区居民的意见之后才开始规划的。这个海港代表着河流和大海的过渡，构成了一个不断漂浮的移动的潮际景观，为市民提供了丰富的亲身体验。海港的设计包含了新的海滩、海滩甲板、一个石头海岬、一个码头、一条木板路、水剧场、游泳区、钓鱼区和划船区。后期的设计包括一个古朴的聚会场所，一个新的停车场和一个咖啡馆。文化公园，艺术品故事展示墙，聚会空间和文艺表演都淋漓尽致地展示了当地土著的欧洲文化特色。整个项目规划旨在创造一个提升市民和游人娱乐品质并保护和提升这个地区的自然美的新的公共空间。

The design work undertaken by ASPECT Studios on stage one of the revitalisation of Jack Evans Boat Harbour is now complete and the new recreational and aquatic playground is open to the public. The upgrade to the Jack Evans Boat Harbour enriches the existing park experience to become a unique civic place of waterfront leisure. The design strongly promotes physical activity by providing a range of public domain elements such as extensive walkways and cycle-ways. Steps, ramps for those with physical challenges, and unique water front conditions allow access to the waters edge for swimming and boating. At the area of 43,000m², the parklands cater for a range of other uses enabled by the promenade including meeting places, weekend markets, memorials, children's play spaces and generous green banks for relaxation. It is also a major tourist attraction for Tweed Heads, on the Queensland and New South Wales border.

The project has been developed to create a diverse, vibrant, culturally rich, recreational and tourism centrepiece for the Tweed Heads Town Centre. It is an exciting foreshore, parklands redevelopment project which will provide the impetus for critical economic revitalisation of Tweed Heads. This project is the culmination of extensive community consultation and thorough planning. The park marks the transition between the river and the sea, framing the ever-shifting inter-tidal zone in a landscape which is in a constant state of change. The design offers a richness by creating a series of distinct experiences with the water——a new beach, beach deck, a rocky headland, urban pier, boardwalk, water amphitheatre, swimming areas, fishing points and boating. Further stages to the design will include an aboriginal meeting place, upgraded parklands and a cafe. Cultural gardens, artwork story walls and space for public, community and performance art will showcase the regions rich local Aboriginal and European heritage. The redevelopment will create a public space conducive to increased local and visitor recreational use, whilst protecting and promoting the natural beauty and environment of the area.

Square & Park

广场和公园

Elwood 海滩
Elwood Foreshore

项目档案

设计：ASPECT 工作室
项目地点：维多利亚，埃尔伍德，奥蒙德海滩
面积：10 000 平方米

Project Facts

Landscape Architecture：ASPECT Studio
Location：Ormond Esplanade, Elwood, Victoria
Site Area：10,000 m^2

ASPECT 工作室和 City of Port Philip 一起完成了海滩的重建工程。他们为当地的居民和周围的社区创造了一个在空间上划分严谨和功能完整的海滩。

像其他海滩一样，这个海滩的公共空间被不合适的停车场、社区、划船俱乐部，还有不合适的排水系统和不充足的树木所毁坏；海滩公共空间和海滩以及海之间的关系被错误定位，汽车和船阻碍了行人的自由行走，同时也妨碍了太阳浴，野餐和骑单车，并且影响了整个地区的景色。而新的设计改变了这种关系，将行人的便利性放在首位，将交通集中在区域的后面，共享的通道减少了单车、行人、船和机动车辆的冲突。简单的混凝土结构，木制表面，当地的植被，整个设计都充满了当地的特色。所有的设施和结构都是水平的，并且造型简单，将对海滩景色影响降到最低。停车场和水箱都利用了节水系统，缓和了清洗船只所需大量水的压力。

这个项目证明了功能和美观是可以无间隙融合的。项目为社区带来了一个公共的吸引人的海滩，同时项目也可作为举行日常大型聚会，航海俱乐部盛会和急救的场所。

Square & Park

广场和公园

Square & Park

ASPECT Studio in conjunction with the City of Port Philip completed the re-development of the Elwood coastal foreshore for local residents and the wider community. Creating simple clarity of space and usability was an essential component in the redevelopment of the foreshore.

This site like many foreshores and their public spaces had been wilted away by inappropriate car parks, community and boat clubs exceeding their limits, inappropriate drainage and lack of tree cover. As a consequence, the primary relationship between the foreshore public spaces and the beach and sea were dislocated with the car and boats dominating the freedom of pedestrians, sun bathers, picnickers, bike riders and views. The design changed this relationship placing the pedestrian as priority and directing traffic towards the back of the site. Shared pathways helped to reduce conflict between bicycles, pedestrians, boats and motor vehicles.

The design draws on more robust coastal vernacular of the region, simple concrete forms, timber surfaces and indigenous vegetation. All new furniture and structures are low, horizontal and simple to underscore the wide bay water views. Water sensitive urban design is used throughout the carpark and water storage tanks are provided to reduce the pressure on mains water for boat washing.

This project demonstrates that a seamless connection can be made between functional design (cars, bikes, walkers, boats, cleaners) and design elegance. The design has provided back to the community an open and inviting foreshore place, which can be used from causal to large scale, surf life saving and sailing club festivals.

广场和公园

Pirrama 公园
Pirrama Park

项目档案

设计：ASPECT 工作室，Hill Thalis Architecture + Urban Projects, CAB Consulting
项目地点：澳大利亚新南威尔士，悉尼
完成时间：2010
面积：1 800 平方米
所得奖项：2010 AILA 新南威尔士奖
　　　　　2010 AIA 国家奖
　　　　　2010 国家儿童安全游乐空间设计奖
　　　　　2010 AILA 国家奖

Project Facts

Design：ASPECT Studio，Hill Thalis Architecture + Urban Projects, CAB Consulting
Location：Former Water Police Site, Pirrama Rd, Pyrmont, Sydney, NSW Australia
Year：2010
Site Area：1,800 m²
Awards：2010 AILA NSW Awards——The Medal1
　　　　2010 AIA National Awards——Walter Burley Griffin Award for Urban Design
　　　　2010 Kidsafe National Playspace Design Awards——Public Playspaces
　　　　-District Parks collaboration withFiona Robb——Landscape Architects.
　　　　2010 AILA National Awards——Planning

这是一个公共空间项目，项目重新诠释了城市中水的输出和引进，水岸社区设施和可持续性设施（包括太阳能和雨水利用）的强大关系。这个项目将原本公共的混凝土路板变成了一个多姿多彩的水岸公共空间。悉尼的砂岩地质构成了世界上著名的海岬，悬崖和海港景观。这个项目并不像其他一些海港公园利用现有的地形还有砂岩，因为这里的地形是平坦的，没有太多特色。设计师利用了 1~2 米长的混凝土平板来铺设整个场地。这个项目最大的挑战就是融合并和谐地处理原有的和现在的环境。

Square & Park

045

广场和公园

这个受到高度评价的项目提供了各种海港娱乐区域。结构上主要由一系列公园房屋和多条新的通道构成。海岸周围的散步道是整个公园最重要的一部分,连接着长达14公里的公共空间网络。海港边的船舶工程构建了一个可以遮风挡雨的海湾,展现了这个地方原有海岸线的风采。公园包括一个新的公共广场,四周有砂岩悬崖、华盖、凉亭和锯齿状海湾。这个充满阳光的公园功能多样,既可用于文化活动和表演,也可以用于会议,交易和节日。

这个综合性公园迅速成为当地人和游客的兴趣焦点。嬉水设施,回收利用的砂岩和天然座位成功地吸引了当地越来越多的居民前来游玩。

项目提供了多条通道和一条水岸散步道。散步道连接着长达14公里的公共空间网络,重新建立起哈里斯街和海港之间的直接联系,创造了一个新的公共空间。

船舶工程构建了一个可以遮风挡雨的海湾,展现了这个地方原有海岸线的风采,同时也巩固了这个区域和悉尼海港的历史关系。一系列公园房屋也因这个独一无二的地理位置而诞生。

Square & Park

广场和公园

Square & Park

Pirrama Park, on Sydney's waterfront is a transformative urban parkland project born from community action. It restores the powerful relationship of the city to the water and delivers unique waterside community facilities, innovative sustainable components including solar energy and far reaching storm water (WSUD) initiatives. Pirrama Park transforms a fenced-off post industrial concrete slab into a richly varied urban, waterfront parkland. The sandstone geology of Sydney creates its world renowned headlands, cliffs and harbour landscape which offer sublime views to and from the harbour. Unlike other recent harbour parks which use topography and the drama of sandstone to great effect, Pirrama Park has a flat, tectonic, and featureless base. The designers inherit a slab of concrete with barely 1-2 metres of variation across the site. The challenge of the design process is to craft a response through a reading of the site both past and present.

The highly awarded design provides a variety of places for harbour side enjoyment. It is structured by a series of park rooms and multiple new paths. The waterfront promenade around the foreshore is the backbone to the park, and an important link in the 14km network of open space. Significant marine engineering at the harbour edge create a sheltered bay and reveals the former shoreline of the site. The park creates recreation opportunities along the waterfront, offering the opportunity to step down and engage with the water. It features a new public square at the waters' edge, defined by the sandstone cliff on the curve of Pirrama Road, the canopy, pavilion and indented bay. This sunny, sheltered place can accommodate a range of public uses including cultural events and performances, meetings, markets, festivals and the like, appropriate to the evolving urbanity of the area.

The integrated site and specific play space have quickly become an attraction for both locals and district visitors alike. The water play elements, reclaimed sandstone and natural setting have successfully catered for a growing population of families in the inner city.

The design provides multiple new paths and a waterfront promenade around the Pyrmont foreshore, an important link in the 14km network of open space extending from Glebe to Rushcutters Bay. It re-instates the direct historic relationship between Harris Street and the harbour and creates a new public square at the waters edge.

Significant marine engineering is proposed to create a sheltered bay and reveal the former shoreline, to create passive recreation opportunities at the water's edge and strengthen the site's historic relationship to Sydney Harbour. A range of other park rooms are created which celebrate this unique location.

广场和公园

蓝花楹广场
Jacaranda Square

项目档案

设计：ASPECT 工作室，MWA，Deuce Design
项目地点：悉尼奥林匹克公园
面积：4 000 平方米

Project Facts

Design：ASPECT Studio, MWA, Deuce Design
Location：Sydney Olympic Park, Sydney
Site Area：4,000 m^2

"每一天体育馆"是悉尼奥林匹克公园第一个新建的公共空间，它再次将奥林匹克的传奇以及奇观聚焦在这个可持续性发展的城镇中心。项目名称一语双关，既认可了 2000 年奥林匹克运动会，同时也精确地概括了这个项目的设计理念。一个典型的体育馆的形态通常都是采用常见的形状，例如，由直角边的座位和墙体构成的站台，组合成一个大型的不正式的开放空间。和体育馆建筑和奥林匹克公园北面的空间相比，蓝花楹广场是一个长和宽各为 80 米和 50 米的长方形，位于火车站的中轴线处，周边有很多新的贸易建筑。这是一个为市民提供娱乐和聚会的地方。在华盖下面有一个圆形的咖啡馆，为市民提供了更为方便的设施。周围的墙体中间采用混凝土，边缘部分采用光滑的砖块。一次性混凝土浇筑的凉廊长椅座位为市民提供了休息的绝佳地方。座位是由两个模具叠加在一起构成的，总长为 155 米，不仅具有娱乐性，深受市民喜爱，而且丰富了广场的边缘设计。

广场的边缘由一个低矮的石造墙体构成，以长方形的形状向周边的街道延伸。从街道望去，光滑的绿色面砖以特定的形式交错在一起，墙体极其漂亮。整个项目中大量使用了混凝土和砖块。因为混凝土便于形成复杂的曲线，而砖块给人的触觉感受是独一无二的，而且具有丰富的文化韵味。室内硬景观大部分也是由彩色的光滑砖块构成的。

这个项目成功地将景观设计，工业设计和图像设计与建筑融合在一起，创造了一个凝聚智慧和具有纪念意义的公共空间。因此，这个项目获得很多奖项，包括 2009 年 Think Brick Horbury Hunt 大奖和 2009 年澳大利亚建筑协会奖。

Square & Park

广场和公园

The Everyday Stadium is the first new public space in Sydney Olympic Park. It refocuses the Olympic legacy of event and spectacle into a finely grained, sustainable town centre place. The project title "The Everyday Stadium" is both an ironic nod to the 2000 Olympic Games and a precise description of the design concept. The morphology of a typical stadium gives the scheme its enduring diagram ie. an orthogonal edge of seats and walls (the stands), framing a large informal open space (the field), and protected by long perimeters of shade ——one built and the other trees (the canopies).

In contrast to the stadium buildings and northern spaces of Olympic Park, Jacaranda Square is a mid-scaled rectangular space of 50 x 80m, situated on the train station axis and framed on its long sides by new commercial buildings which define it as the pivotal public space of the new town centre. The park is a place for passive recreation and a centre for community gathering. A circular cafe situated beneath the canopy provides an additional level of amenity for the new community of workers. The perimeter walls, which are composed of glazed brick to the street and concrete to the centre, have been designed to provide joy and amenity along their entire length. One-off precast seating made up of a series of modular precast concrete provides lounge suite resting place for everybody. The seats are made up of two moulds and are designed to "telescope" and step along the 155 lineal metres of perimeter wall. They have a playful and personable scale and make an active and occupied edge to the inside lining of the square.

The edge is defined by a low masonry wall set out orthogonally to the surrounding street grid giving the space a strong rectangular form. From the street, the wall appears as a rich and dramatic frame, laid out in a finely grained "brixel pattern" of green glazed and face bricks that define the edges and entries to the park. Concrete and brick are used extensively in the project, concrete for its ability to achieve complex curves, and brick for its fine grained tactile quality and for its cultural, place making associations to the adjacent Homebush Brickpit. As well as the polychrome glazed bricks, the majority of the interior landscape is made up of recycled bricks.

This project successfully fuses landscape architecture, industrial and graphic design with architecture to create an intelligent and memorable open space. It has received many awards including the 2009 Think Brick Horbury Hunt Award, 2009 Australian Institute of Architects (NSW) Award, Commendation in Urban Design 2008 MBA NSW Awards Civil Engineering.

广场和公园

717 Bourke 街道
717 Bourke Street

项目档案	Project Facts
设计：ASPECT 工作室，Metier3 Architects	Design：ASPECT studio, Metier3 Architects
项目地点：维多利亚港口	Location：Docklands, Victoria
完成时间：2010	Year：2010

这个项目位于两座建筑之间，设计具有很大的挑战性，而最终构建的多层次的城市景观呈现的是一份满意的答卷。

ASPECT 工作室参与了街景、街心花园、演讲前院和后院的设计。在 Metier3 建筑事务所和 PDS 集团的协助下，一个新的区域被加进整个项目。为了打破景观和建筑典型的 90 度关系，ASPECT 工作室创造了一种新的景观构成形式语言，设计了坡道、座位、园圃和露天平台。

Square & Park

A multi-level urban landscape realm wrapped around two buildings drove the design towards a highly architectural outcome. ASPECT Studio was engaged to design and document the streetscape, pocket park, podium forecourt and courtyards for this mixed use development in the Docklands. Working collaboratively with Metier3 architects and PDS Group, ASPECT studio had added a new precinct to the Docklands. Wanting to remove the typical 90 degree relationship with landscape and the building, ASPECT Studio created a language of tectonic landscape forms, made of ramps, seats, garden beds and decks.

广场和公园

社区公园
Neighborhood Park

项目档案

设计：Cino Zucchi Architects
项目地点：意大利，威尼斯，San Dona di Piave
面积：20 000 平方米

Project Facts

Landscape Architecture：Cino Zucchi Architects
Location：San Dona di Piave, Venice, Italy
Site Area：20,000 m²

这个项目是 San Dona di Piave 的一个雪白色的社区公园。San Dona di Piave 是威尼托区一个典型的中等大小的城镇。这里是一个大型的工业和商业区域，唯一的景观便是阻止洪水的大坝，商业区公园和一个相当普通的住宅区，户外空间设计非常差。

这个项目的设计目的在于创造一个小型的社区公园，为居民提供一个现代的公共空间。不合理的街道社区规划，普通的房屋建造结构，极少的公共服务设施，不充足的公共通道，这些都需要重新设计。一系列二至四层楼高的房屋被重新改造成坐北朝南，面向主要街道。在主要街道旁边有一个停车场，停车场呈90度角形成一个豆状的表面结构，并用沥青铺设，以沙丘点缀，看起来就像是一个天然的河坝。人行道由混凝土构成，表面上用小的白色卵石装饰。不同的人行道有不同的宽度，形状和层次，所以也就具有不同的功能：座位、户外剧场、喷泉、儿童聚会区、单车道、野餐区等等。一个木制屏风以螺旋状围绕成一个游乐场，照明的灯看起来像是破土而出的豆芽。在较高斜坡的凹处喷出一股股水，水顺着坡流下，构成了一个不正式的露天剧场。中间十字形的凹处沿着坡道而下，越来越宽，可以方便坐轮椅的游客。一个人工的源泉位于一个稍高的山丘之上，源泉溢出来的水的体积大小和高度会在半个小时之内交替改变，流出来的水集中在一个月亮形的池塘里。

Square & Park

广场和公园

树木，灯光设施和白色的石头长椅不仅构成了空间，而且在不同的季节里创造出不同的微气候。喷泉旁边的中心区域被大面积的树木遮盖着，在夏天的时候成为清凉的一角。一棵挺拔的橡树下围绕着一圈圆形的混凝土长椅，人们可以坐在这里聊天或野餐。土丘的高度经过精密计算，不仅没有妨碍到中间区域大面积的草坪，而且也没有挡住观望房子的视线。放射状的通道将中心区域连接着单车道和人行道，并且延伸到城市中去。整个项目整体性和连续性非常强，不是分离开的一个个独立的服务区域来满足有限的娱乐需要，而是减少视觉与空间上的不和谐性。

在阳光最强烈的时候，白色表面反射出来的光尤其刺眼，所有社区的人也称这个公园为"雪白公园"或"日光浴公园"；在不同的时节，不同的时间，白色表面会反射出不同的颜色，甚是好看。这不仅在景观内诠释了自然和天气变化，而且在日常的城市生活中增添了很多自然之美。设计的夜灯同样给公园增色不少。一系列细长的灯杆沿着曲线的通道隔段设置；依附在地面上的灯更显出地面的粗糙之美；较高的灯杆将公园与周边的公路区分开来。

这个公园在任何时候都是一个有活力的聚会地方，人们可以在这里读报纸，骑山地车，聊天和弹吉他。

广场和公园

This project is a snow-white neighborhood park in San Dona di Piave. San Dona de Piave is a typical medium-sized town in the Veneto region, which works as a large expanded industrial and commercial area, in a landscape alternating the earth dams protecting from the Piave river floods, retail parks, and a rather mediocre suburban mid-density sprawl of residential types with poor open space design.

Square & Park

The public commission by the city of San Dona de Piave to design a small neighbourhood park located in the middle of one of these suburban developments was the occasion to experiment on the theme of a contemporary public space in this middle landscape. The initial datas were somehow discouraging: a poor design of the new street network, mediocre architectural solution for the houses, little or no public services to animate the new space, strongly residual character of the plot assigned to the park, public accessibility. The shape of the given plot surrounded by two-to-four floors high houses has been redesigned to open it up to the south toward the main street. Next to this main access, the planned parking has been turned ninety degrees to form a beanshaped surface paved in brightly coloured asphalt and screened by an earth dune recalling the river dams of the context. A main hard surface of small white pebbles cast in white concrete unifies the pedestrian paths, deforming its width, shape and level to become in turn, a seat, an open-air auditorium, a fountain, a gathering place for kids, a bicycle route, a picnic spot. A spiralling wooden screen shelters a playground lit by a high lamp pole in the shape of a sprout. A series of ripples in the concave higher slope acts as an informal amphitheatre for events, and their central crossed height offset create a low-pitched ramp for wheelchairs. An artificial water spring with time-planned intensity differences in the strength and volume of the stream at half-hour intervals cascades down from the higher mound through a series of slits to a moon-shaped pool.

广场和公园

Square & Park

Trees, lighting fixtures and white stone benches pierced by tree pots help define the space and create different microclimates in the different seasons. A central area next to a drinking water fountain is shaded by denser trees, offering a fresher spot in warm summers; a roughly circular concrete bench is bent around a single large oak tree to seat informal chats or picnics. The height of the earth mounds which embrace the wide central meadow is studied to protect the space from the surrounding disorder but not to screen off the view from and to the houses. A number of radial paths connect the central space to the bike and pedestrian network which connects the neighbourhood to the larger city structure. A continuous earthwork, rather than a number of separated services, respond to the few needs of well-being and leisure of a park, contributing to reducing rather than increasing the visual and spatial disorder of the suburban sprawl.

The glaze from the paying, quite strong in the sunniest days (the neighbours tenderly label it Snowwhite or the Suntan park) slowly changes into coloured shades at different hours and seasons, amplifying the perception of natural events and weather changes as a large sundial or a landscape, reconciling the natural dimension with the daily city life. The evening light design amplifies the different character of the areas: an array of slender light poles follow the gentle curve of the paths, ground-level horizontal light points exaggerate the roughness of the paying surface, and the high light pole in the playground signals the park from the perimeter road.

The park has already become a very lively meeting place at all hours, hosting relaxed newspaper readers as well as mountain bike acrobats, old ladies gossiping and youngsters playing guitar.

广场和公园

Banyoles 公共空间
Public Spaces in Banyoles

项目档案

设计：Josep Mias Gifre, Mias Arquitectes
项目地点：西班牙，赫罗纳，Banyoles
面积：18 000 平方米

Project Facts

Landscape Architecture：Josep Mias Gifre, Mias Arquitectes
Location：Banyoles, Girona, Spain
Site Area：18,000 m²

Banyoles 这个城镇的环境曾经很恶劣。狭窄的街道，古老的人行道，车辆和行人都杂乱无章地存在这样一个混沌的城镇系统中。原本清澈的水渠成了肮脏的下水道。中心广场的人行道上也到处停了很多私家车。项目的目标就是清除这个区域的所有车辆，使之成为行人专用区。这个项目中主要采用石灰华石头。这种石灰华石头主要存在于城市中的下层土中，所有的神秘建筑，教堂和中世纪住宅或纪念馆都是用石灰华建造的。

中心广场是这个项目的主要部分。用石灰华以镶嵌的形式铺设整个广场。这种铺设方法在项目街道修建和小广场等方面也都有利用。另外一方面，灌溉系统沿着人行道间歇式地敞开着，这样就给孩子们带来嬉水的乐趣。

通过重新铺设中心广场来创造一个新的行人专用区，这也和 Banyoles 中世纪遗留下来的一部分一致。项目的第二部分主要是恢复灌溉水渠。实际上，这些水渠源于 Banyoles 的一个池塘，经由城市，流向居民的后院。私人花园的消失让这些水渠渐渐失去了它们在城市的功能性。

这个项目恢复了这个城镇的人与水的流畅性。材料的选择与城市中心保持一致。在设计上，人行道突破以往的线型结构，注重道路表面的不完整性与层次性，让表面有一种水面的感觉。

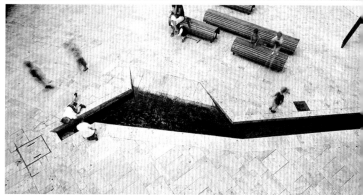

Square & Park

广场和公园

Square & Park

Banyoles' old town used to be a deteriorated area in which vehicles and pedestrians cohabitated around a urban system of narrow streets and old sidewalks. The irrigation canals that originally were clean had become part of the sewer system of the city. Around the Central Square there were also sidewalks in which cars parked randomly. The process was to pedestrianize the whole area, and remove all the old sidewalks. The new intervention is made with travertine stone. This calcareous stone has always been present in the city's subsoil. All the enigmatic buildings, churches, medieval houses or monuments were also raised with travertine.

The departing point is to cover the Central Square (the most relevant part of the project) using a tessellation of travertine. The proliferation of this tiling arrives to the streets and minor squares in different phases of the project. On the other hand, the irrigation system is uncovered intermittently across the pedestrian ways. Eventually, it is opened in bigger sections so children can play as if they were in front of a puddle of water.

Re-paving the city center defines a new pedestrian area. It corresponds to a part of Banyoles in which the traces of the medieval age are still present. The second part of the intervention was to recover the irrigation canals. Originally, these waterways came from Banyoles' Pond, crossed the city and supplied water to the backyards of the houses. The disappearance of these private gardens generated the progressive covering and degeneration of the quality of the canalized water.

The project restores circulation of people and water through the old town of Banyoles, giving them back the itineraries they occupied originally. The designers chose the same material by which all the city center is built. The tiles of travertine stone generate folds in order to form canals or regulation gates. We break the lineality of the pedestrian paths making cuts in their surface so the flow of water can be felt. Certainly, the purpose is to exhaust the possibilities of the material of travertine itself, from the soil to the new paving passing through water.

Rubi——市政市场大楼广场
Rubi, Municipal Market Building and Plaza

项目档案

项目地点：巴塞罗那, Rubi
面积：1 600 平方米（办公）
　　　5 600 平方米（停车场）
　　　2 500 平方米（公共空间）

Project Facts

Location：Rubi, Barcelona
Site Area：1,600 m² (offices)
　　　　　5,600 m² (parking)
　　　　　2,500 m² (public space)

这个项目位于城市中心位置。因为地块呈三角形，所以市场也是三角形的。在北面，起初是一个空地。因此就将这一部分设计成一个地下两层的停车场，并和原有的停车场相连接。在新的停车场上方，市场也因此得到扩建。项目包括底层楼面的整修，冷藏室和垃圾处理站的建造。新的广场就是市场新的户外大厅。在地形上，着重斜坡的可行走性以及合适性。另外，这个广场也延续了市场内的贸易活动，因此有必要为交易和市场的贸易活动留出足够的空间。

这个项目改变了市场原来的方向，将主要入口设计在广场内。所以，扩建部分像是一个建筑立面，既有新的城市大厅，也有市场行政办公室。这座建筑位于市场的前面，并且比市场要高出些许，看起来像是一个公布牌。在大楼的一层，贸易空间被预留出来，另外市场的主要入口和地下停车场也有设计。

项目的设计想法不仅要考虑到场所，而且还要考虑到在这个场所进行的活动。有些想法是在紧张的工作过程中诞生的，但往往这些想法对项目的顺利完成很重要。

这个项目可以成为"城市海滩"。在这里你们可以躺着享受日光浴，可以散步，也可以展开娱乐活动。沙子和人行道构成了新的地形。大楼可以被称为"水建筑"。透明的蓝色建筑表皮，水泡沫依稀可见。景观上凸显海岸风格，主要采用蓝色和灰色。

Rubi local market is in a central urban position. It is in a triangular plot, which gives the market the same shape. In the northern side, before the intervention, there was an excavated unbuilt and residual space which was not used. The project proposes building an underground two-storeyed parking, connected to the existing one. Over the new parking, the enlargement of the market is built. The project includes the refurbishment of the lower floors in order to integrate them into the parking, installing refrigerators and waste processors. This new square, over the underground parking, will be the new outdoors hall of the market. Its geometry tries to make the slope walkable and suitable as a public space. In addition, it is an extension of the commercial activities inside the market, giving enough space for fairs and occasional outdoor commercial events.

With this project, the market changes its orientation, having the main entrance in the square. So then, this extension is almost like a facade-building, with new city hall and market administration offices, which lies on the northern side of the plot. This building then occupies the frontal area, growing higher than the existing market so that it becomes something like a three-level placard. In the ground floor, the building has commercial spaces related to the existent market activity, and the main access to the market and to the underground parking.

In fact, projects have always a secret thinking or a hidden beginning which allows thinking about the place and working in it and this is something which is normally not explained. They are personal interests, certain ideas that were born in an intense work moment, which happen to be important for the logical development of the project.

We have built a sort of urban beach, where people would lie and sunbathe, wander, play. It is a new topography made of sand and pavement that landmark a water building, something like a wavy front. This is a transparent blue building, where foam is still shining. It is almost a marine landscape made of blue and grey, like the roof of the market, where water would be the main character too.

Square & Park

哈格里夫斯购物广场
Grounds of Hargreaves Mall

项目档案

设计：Rush Wright
项目地点：澳大利亚，维多利亚，本迪戈
完成时间：2010

Project Facts

Landscape Architecture：Rush Wright
Location：Bendigo, Victoria, Australia
Year：2010

哈格里夫斯购物广场成为本迪戈中心新的公共空间。在去年甚至过去更长的时间里，这个城市在CBD战略基础上，开始着手建设一个集商业和活动为一体的公共空间。这个项目将原有的零售中心转变为一个有活力的、功能齐全的区域。这个区域不仅包括市民所需要的日常生活需求以及商业贸易功能，而且包括多层和商店上面的住宅。在不久的将来，这里会成为人们居住，工作和休闲最重要的空间。整个项目充满了争议，当地有很多私家机动车的爱好者，这个项目无疑影响到他们。但是，最后这里成为哈格里夫斯的中心。扩建的区域是没有路边石的，在这里汽车，单车和行人都有各自的空间。

哈格里夫斯购物广场原本只是一条单调的街道，在20世纪80年代购物广场盛行的时候，原本的街道才被改建成购物广场。原来的设计是将街道改造成半封闭的广场，并包含当时所有的流行元素。考虑到曾经的交通和传统的街道地形，以及将来的居住环境，这个方案被客户和设计团队所否定。

街道的本质就是在每个层面上呈现新的设计。最终设计的结果是：一排排树木，步行区域，车辆，单车区域以及其他交通设施。进出口是由液压缆柱控制的。

客户要求设计师设计两个男女可以通用的洗手间和多个报刊亭设施。另外，设计师还设计了婴儿换尿布处，看上去像是一座塔；而这样的创意设计主要来自婴儿的头部和当地的一些设施。塔的下面悬挂着可以拆除和移动的遮阳板，另外还有很多悬臂式的玻璃屏幕。

这个屏幕同时也被用来展示各种图片。特别设计的玻璃屏幕能够恰到好处地反射阳光，同时变成了一个投影屏。照明灯的半透明性和白天时候的美观性，以及投影屏幕的大小和高度等这些问题都得到了很好的解决。在这里，市民可以观看世界杯或者是本迪戈美术馆的作品以及当地工匠的作品。

这个项目意在为这个城市创造一个长久的持续的传奇。客户和设计团队对于这个都有很高的要求。这个项目与这个城市自从19世纪以来的重大发展紧密结合，代表着高质量的核心项目，刺激了这个城市的未来发展。

Square & Park

Hargreaves Mall forms the new centrepiece of the public domain of central Bendigo. Over the last seven or more years, the city has embarked on an award-winning redesign of its central commercial and activity areas within the strategic context of the CBD plan adopted by Council. This will see the CBD transformed from a predominantly retail core into a vibrant, mixed use area combining the full suite of civic and commercial urban functions with new types of urban housing including multi-level and shop-top housing. In future it is envisaged that Hargreaves Mall will become a form of ramblas with people living, working and spending much of their day and night in the near vicinity. The Mall already forms the heart of the new "Walking Bendigo" pedestrianisation, a controversial project locally especially amongst diehard users of private motor cars. Extensive areas of kerb-free shared zones are being created where cars, bikes, and pedestrians negotiate for space.

Hargreaves Mall used to be Hargreaves Street, and was transformed into a pedestrian mall during the flurry of mall building in the 1980s. The old design attempted to turn the street into a semi-enclosed plaza, replete with one of everything from the repertoire of the time. This is best left to the jurors imagination. Consideration was of course given to restoring vehicle traffic and a completely conventional street typology as has occurred elsewhere, but in light of the Walking Bendigo project and the potential for future living environments right on the mall, this was rejected early by the client and design team alike. The essence of a street, however, informs the new design at every level. The designers have restored a conventional street morphology of tree rows, walkway zones and clear pathways for vehicles, bikes and other occasion service traffic. Access is controlled by hydraulic bollards.

广场和公园

Square & Park

广场和公园

Square & Park

The client required the design team to include two unisex toilets and provision for retail kiosks. These have been presented together as large scale civic marker structures. These forms are derived from the minimum plan dimension of a unisex toilet and baby change, then projected upwards as a tower form, evocative of poppet heads and local mining infrastructure. From the towers are hung both retractable sun shades for the commercial space beneath, as well as dramatic cantilevered glass screens.

The screens have also been designed to display projected images. The digiglass interlayer chosen through extensive prototyping also has the ability to reflect just the right amount of light to become a projection surface. A careful balance was struck between the competing requirements of lantern translucency, daylight appearance and suitability for projection of video and other moving or static images. A Grand Final or World Cup could work well projected onto the screens, as could art-based imagery from Bendigo Gallery and local artisans.

This new mall for Bendigo aims to provide a very long lasting, durable legacy for the city. The bar of ambition, by client and design team, has been set high. The new space sits well alongside the many civic achievements in Bendigo most of which date from the 19th century. A new age of city building has begun, with the new Hargreaves Mall as a nucleus of quality and catalyst for future development in the CBD.

广场和公园

Rooke 自然保护区公园
Rooke Reserve Park

项目档案

设计：CPG AUSTRALIA
项目地点：澳大利亚，维多利亚
完成时间：2010
面积：15 000 平方米

Project Facts

Landscape Architecture：CPG AUSTRALIA
Location：Victoria, Australia
Year：2010
Site Area：15,000 m²

这个项目位于墨尔本以西20公里的一个住宅区里。绝佳的地理位置和独一无二的设计，使得公园的行人道和单车道紧密地与周围的住宅社区，学校和幼儿园，公共空间连接在一起。这个公园主要为人们提供游玩场所和设施。设计师综合考察这个区域的地质，地形和材料，将不同的游乐设施设计在不同的区域，功能性明显被区分开，更方便人们，特别是2至15岁的孩子们使用。

这个公园为人们提供了个人和团体游玩设施。足球，滑梯和秋千都比较适合团体使用。而玄武岩上的字母和枯竭的小溪都会吸引个人去探索。游玩和水紧密结合在一起。水经由加高的草坪和花圃流入底层的花圃中，最后流进地下的混凝土水槽；这些水又可以被重新利用。在玄武岩平原上，突兀的花岗岩岩层，成为另一种别致的景观。零碎的玄武岩构成干涸的小河的河床，让人不禁联想起附近的一条河流，或者西部平原农场的栅栏。

Square & Park

Rooke Reserve is a new parkland centrally located within a new residential development in Truganina, approximately 20km west of the Melbourne CBD. Its location and design provide clearly defined pedestrian and bicycle connection to surrounding residential neighbourhoods.

Play is central to the design intent of Rooke Reserve, with both structure and setting designed to facilitate the specific requirements for 2-5-year play, 5-12-year play and access for all play. Whilst structured play elements highlight specific play zones, they also act to trigger investigation of the sites texture, topography and materials.

The park design provides for the group and the individual. Whilst steel soccer goals, dual carriage slides and multi-person swings all encourage group play, there are moments within the park such as the basalt letters and the dry creek trail which allow the individual to explore and exist on their own.

Play is also cross programmed with water sensitive urban design elements such as the dry stone creek trail. The water runs off from elevated lawn and garden beds to precast concrete water tanks (underground) for irrigation re-use. The texture and topography of the Western Basalt Plains provide a narrative throughout many of the design elements of Rooke Reserve, with direct references made to the You Yangs, a steep granite outcrop rising abruptly above the contrasting Werribee basalt plains. Naturally occurring basalt "floaters" uncovered during excavation and construction of the park and surrounding estate, were retained and reincorporated as the diy creek bed, reminiscent of neighbouring Skeleton Creek, and worked into the extensive dry stone walling, creating a link to the historic farming fences of the western plains.

广场和公园

Nicolai Kultur 中心
Nicolai Kultur Center

项目档案	Project Facts
设计：Kristine Jensen	Landscape Architecture：Kristine Jensen
项目地点：丹麦，科尔丁	Location：Kolding, Denmark
面积：3 400 平方米	Site Area：3,400 m²

Sct Nicolai 学校中心原本同其他传统的学校一样，只有运动室和操场。本项目意在创造一个新的具有创造力和动态的文化中心，共新增加了五个艺术室。这个项目将学校操场作为中心，围绕这个中心构建了功能设施。电影室、文学室、儿童室、手工房、音乐室，一改历史的面貌，成为一个自由和愉悦的空间。

在学校的西边，建造了一堵连续的由柯尔顿耐腐蚀钢铁构成的墙。另外，学校操场也增加了不同的户外活动空间——故事园、露台、城市园、后院和广场。操场的表面倾斜，由沥青铺设，并且点缀了很多由热塑性塑料和油漆构成的图样。这些图样都是白色的，与黑色沥青形成鲜明的对比。同时，黑色的沥青表面也可以被用作黑板，以供孩子们在地面上尽情作画。

Square & Park

这个项目的目标在于将整个区域转变成一个文化综合体，不同的部分有不同的功能。例如，户外的舞台设计为孩子们提供了表演和欣赏的地方；探索与交流区和孩子们的玩耍区——区别开来，各具功能。设计的目的在于运用新的设计想法，创造一个现代的学校操场，让学校变得更加具有活力性、娱乐性、教育性和参与性。因此，在学校的一边增添了一堵长达 70 米的墙体，在另外一边为原有的房子增加了新的结构。新建的墙体是由柯尔顿耐腐蚀钢铁构成的，呈橘红色，连接着学校各种各样的建筑。除了白色和黑色，还有蓝色，黄色和红色；这些颜色主要构成了沙盒和水池。

广场和公园

Sct Nicolai School Centre is a continuous reincarnation of everything from the traditional Danish school with gyms and schoolyards. With five art houses as the new function, the possibility of a new typology of creative and dynamic culture centre emerges. The project evolves with the idea of the schoolyard as the centre, joining the different functions together, an active free space in a more cheerful schoolyard than the historical version. A yard creates a base of the different house programs, movies, literature, house of children, craftsmanship and music.
Along the west side a continuous wall of corten steel is built. Furthermore different outdoor activity spaces are added to the schoolyard: The Garden of Tales, The Terrace, The City Garden, The Yard and The Square, opening up towards Katrinegade. The tilting terrain in the schoolyard is prevailed and a new enclosing surface of black asphalt is laid out, on which patterns are drawn with thermo plastic and street paint. The schoolyard is designed with a big white drawing on a new layer of black asphalt as a new drawing board. The project development has been theme based and five groups of users have during the design process participated in concretizing the programming and the future schedule of utilization and maintenance.

The aim for the site has been to turn the whole area into a cultural complex, each house having different purposes as follows: House of literature, House of films, House of crafts, House of children's culture and House of music and the schoolyard should be created so that it would benefit all of the dynamic features from the houses. The designers harvest following programs: a cinematic place as an outdoor scene also able to feature small concerts or plays, places for exposition and markets, places for children to play and everyone else to stay gather, run and just be in the sun. The aims of intervention by an architectural point of view are as below: On the city level or structural codex the designers did two things to full fill the yard on both sides; a new wall almost 70 meters long and a new structure in the row of houses on the other side. The designers are to make a new concept for a modern school yard, which is a lot livelier, full of joy, participation and through feeling of dynamic purpose.
The corten steel creates, with its orange colour, an inclosing wall that links the various buildings together, but also creates areas for recreation like the terrace and the stage. The blue, yellow and red in the garden are elements for playing such as a sandbox and water basins.

广场和公园

Choorstraat 庭院
Choorstraat Papenhulst

项目档案	Project Facts
设计：Buro Lubbers	Landscape Architecture：Buro Lubbers
项目地点：荷兰，斯海尔托亨博斯	Location：S-Hertogenbosch, Netherlands
面积：1 800 平方米	Site Area：1,800 m²

这个项目是由一个修道院改建而成的，只有靠街的那个立面被保留。在这个立面后面，现代的建筑赋予了内部庭院一个全新的面貌。通过拆除掉修道院两边的建筑，在扩大了原有庭院的基础上设计了这个庭院，为整个公寓大楼增添了一道亮丽的风景。

规划项目旨在通过特殊住宅的建造和将私人院落转为半公共广场来提升城市的质量。新建庭院为一系列城市公共空间增色不少，有私人的特点，也很隐蔽，但又是繁闹城市中一个相对开放的空间。人们可以在这里休息、吃东西、听音乐等等。设计简单大方，散发出安静祥和的气氛。

庭院的地面是由烧制的爱尔兰石头铺设的。这些石头呈条状形，看起来杂乱无章，实际上是有规律的。在地面上还有条状木铺设。部分条状木作为下水道的遮盖物，在另外一部分条状木上设计了座位。人们可以选择坐在向阳的一边或者是遮阳的一边。项目设计的焦点是一张长达24米的木制桌子。在桌子中间设计了专门的酒水冷却设施。在这个设施下面连接有管道，并通向地下，是不可见的。另外，24个木制水龙头在桌子上整齐排开，为整个设计增色不少。

庭院是在一个垃圾场地上建造起来的，这样的地理位置限制了设计。特点鲜明的树木树叶非常薄，阳光穿过树叶，不仅没有挡住立面的视线，还在地上和立面上投下斑驳，甚是好看。在晚上，这些树木被地面的灯照亮，营造出一种温馨的氛围。在L形场地的转角点，一些特别的灯杆以不同的高度直立，每一个灯杆都有很多聚光灯，方向各不相同。

A former cloister in the historic centre has been transformed into a luxurious apartment building. Only the front facade in the street has been remained. Behind this facade, contemporary architecture gives the inner court a new image. By demolishing the side wing of the cloister not only the existing court was expanded, also the facade of the neogothical chapel became visible. In this marvellous historic setting combined with modem architecture, Buro Lubbers designed a courtyard that belongs to the sequence of courts and squares around the Sint-Jans-Cathedral in the lively town centre of S-Hertogenbosch. Furthermore, the garden offers a spectacular view at the cathedral.

The starting point for the design was to create a quality boost to the city by the addition of special housing and the transformation of the private courtyard to a semi-public square. The new square would be an attractive addition to the series of urban open spaces in town, such as the market, the squares Kerkplein and Parade, the gardens of the Orangerie and the Casino, and the Vughterdriehoek. The courtyard has a private character——hidden while open spot in the busy city. It provides space to relax, to sit, to dream, to eat, to look, to stay and to listen to the carillon of the Sint-Jans-Cathedral. The modesty of the design created a site that radiates serenity and allure.

The courtyard is paved in an apparent, random pattern with narrow strips of burned Irish stone rich of fossils. Here and there wooden strips are included in the pavement on which banks and chairs are positioned. The banks and chairs are placed so that one can choose between a place in the sun or the shade, under the trees or on the square overlooking the rooftops of St. Jan. The central design element is the wooden table of 24 meters in which a water element is integrated serving as wine cooler. In the table the ducts of the underlying parking are also included, invisible. 24 wooden puff are grouped around the table.

The location of the garden above the garage limited the design possibilities. Where possible —— right next to the existing chapel and next to the monastery garden——characteristic trees are planted in the full ground. The thin leaves of the dead bones tree (Gymnocladus dioicus) shows light, but its transparancy does not cover the facade of the chapel. The trees are lighted from the ground creating a special atmosphere in the evening. From the hinge point of the L-shaped square some special lighting pdes are placed that vary in height. They are important three-dimensional elements in the square. Each pole has a number of spotlights to highlight particular objects.

Square & Park

广场和公园

Malov 轴心流畅景观
Fluent Landscapes / Malov Axis

项目档案

设计：Adept and LiWplanning
项目地点：丹麦，Malov
面积：1 000 平方米
完成时间：2011

Project Facts

Landscape Architecture: Adept and LiWplanning
Location：Malov, Denmark
Site Area: 1,000 m²
Year: 2011

Malov 轴心和毗邻的城市空间项目增添了 Malov 这个城市的活力，促进了城市与周围景观的和谐统一。这里为市民和游客提供了安全的驻足空间和聆听空间故事的机会。

这个项目的主要目的在于将宽阔的景观融入到自然环境中去。受到冰河时代的启发，景观设计主要采用冰碛石。整个地形类似冰川，有冰洞，峡谷和平原。这个项目进出方便，不仅可以通过 Malov 自然公园从北面进入，而且也可以通过 Sodergaard 湖从南面进入。

这个项目流畅性很强，不仅丰富了城市生活和自然环境，还为市民提供了安全的和有趣的生活体验。

Square & Park

广场和公园

Square & Park

Malov Axis and the adjacent city space is a visionary project underlining the lively Malov city character in close contact to the amazing surrounding landscape. The movement trough the axis will, beyond being easy accessible for everyone, safe and dear, provide a spatial story relating to the distinctive character of Malov and create an experience for citizens and visitors.

The main idea sets off in a reading of the area wide landscape and natural context. The landscape is a characteristic moraine landscape shaped through the impact of the glacial period. The glaciers have left a rolling landscape with kept ice holes, valleys and plains. A beautiful and recreational landscape is accessible from north via Malov Nature park and from south via Sodergaard lake and the Malov wedge.

As a result, the Malov axis will appear as a fluent landscape building on the great historical values in the landscape creating a vibrant area full of city life and natural environments, rich on experiences, interesting and safe for everyone.

广场和公园

锡姆科波浪桥
Simcoe Wavedeck

项目档案

设计：West8 + DTAH(joint venture)
项目地点：加拿大，多伦多
面积：650 平方米

Project Facts

Landscape Architecture：West8 + DTAH (joint venture)
Location：Toronto, Canada
Site Area: 650 m²

锡姆科波浪桥是加拿大四大木栈道之一，不仅具有艺术性而且具有功能性。桥梁位于锡姆科大街水边，这个木栈道形状特别，呈波浪形，有大波浪和小波浪。最高的浪峰距水面2.6米。市民可以通过小波浪桥直接通过小河。如果市民想游玩一会，就可以通过大波浪桥。受到加拿大村屋和安大略湖的启发，这个项目的目的在于让市民感觉生活在湖面上。

除了在锡姆科大街水边安装木栈道，这个项目还包括码头岸壁的整修，渔业住宅的建造和景观的提升。这个项目是加拿大滨水区中心项目的第一期，主要由West8 + DTAH 负责设计。尽管这个地区数十年来经历过多次的规划和改建，但是不论是在视觉上还是在功能上，都能与原有的建筑，设施和环境相协调。

The Simcoe WaveDeck, one of four uniquely Canadian wavedecks planned for the area, is as artistic as it is functional. Located just west of Simcoe Street at the waters edge, the wooden wavedeck features an informal public amphitheatre-style space with impressive curves that soar as high as 2.6 metres above the lake. Inspired by the Canadian cottage experience and the shorelines of Ontario's great lakes, the wavedeck is meant to give urban dwellers a feel of life at the lake. Providing access to Lake Ontario is a key priority of Waterfront Toronto.

In addition to the installation of the wavedeck at Simcoe Street, construction activities included dockwall repairs, in-water fisheries habitat construction and landscape improvements. The project is part of the first phase of implementation of a strategic masterplan for the Toronto Central Waterfront prepared by West 8 + DTAH. Yet despite decades of planning and patchwork development projects, there is no coherent vision for linking the pieces into a greater whole visually or physically. In this context, the fundamental objective of the project is to address this deficiency by creating a consistent and legible image for the Central Waterfront, in both architectural and functional terms.

Square & Park

广场和公园

马德里 RIO 景观
Madrid RIO

项目档案	Project Facts
设计：West8	Architect: West8
项目地点：西班牙，马德里	Location：Madrid, Spain
面积：8 000 000 平方米	Site Area：8,000,000 m²

这个项目是马德里市长 Alberto Ruiz-Gallard ó n 一项雄心勃勃的计划，沿着 M30 高速公路，全程长 43 公里，6 个项目段临河，沿途基础设施建设费用高达 60 亿欧元。West8 景观事务所和马德里著名建筑师一起制定了 Madrid RIO 总规划。设计概念是"3 + 30"，一个分三步发展的战略计划。总项目中包含 47 个子项，总造价达到了 2.8 亿欧元。除了各种广场，林荫大道和公园，一系列桥梁也列入了项目当中。这些桥梁用于连接河流两岸的城市区域。

Salon de Pinos 是一个线性的绿色广场，连接着现有的以及新规划的城市空间和河道的带状绿地。广场的绝大部分位于高速路的隧道上方，植物的选择参考了当地山区的树种，主要为松树。通过修整和选择各具特色的树，合理配植从而创造自然化或雕塑化的空间。

Avenida 是通往马德里最主要的道路之一，介于高密度的住宅区和前西班牙国王狩猎场（Casa de Campo）之间。将道路重新置于隧道之中，并且重新设计后将停车空间置于地下，提供了 1000 个车位，释放出地面空间，并将这个空间转变成花园，为当地的居民带来方便。

Avenida 延伸至里斯本，跨越了葡萄牙著名的"樱花谷"，由此抽象的樱花成为公园的一个设计元素。公园种植了不同品种的樱花来延长花期，并延续了葡萄牙的铺装工艺以求与周围空间形成呼应，使得这里成为一个受欢迎的公共空间。

"城市宫殿"具有巴洛克风格，经过合理设计，连接着皇家宫殿以及河对面的狩猎场和果园。基础设施的变迁使得曾经的果园变成了交通枢纽。与最初计划的历史重建相反，设计中采用了现代的手法来阐释传统的果园。在里面种植品种繁多的树木，无花果、杏树、石榴以及类似的象征天堂的树种。同时，管道中的河流又重见天日，蜿蜒于果园中，其源头和河口都经过特殊处理。

Parque de la Arganzuela 以水为主导的场地，主要是亲水空间。公园基于不同的情感和景观，以水为介质，使得这个元素能够被感知和体验。在这里你可以感受到一年四季水流的变化，并观赏水生植物。流经公园的河流会交叉或穿越不同的地形，形成高低错落的空间和不同的主题。

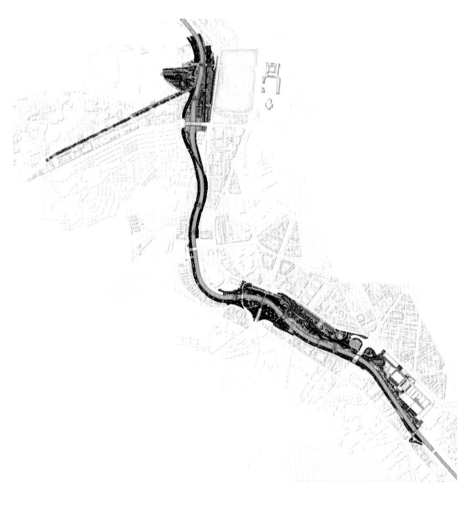

Square & Park

广场和公园

The ambitious plan by Madrid's mayor Alberto Ruiz-Gallardon to submerge a section of the M30 ring motorway immediately adjacent to the old city centre within a tunnel was realised within a single term of office. The city undertook infrastructure measures over a total length of 43 kilometres, six of them, along the banks of the River Manzanares, at a total cost of six billion Euro. West 8 together with a group of renowned architects from Madrid, united under the name MRIO arquitectos led by Gines Garrido Colomero designed the master plan for Madrid RIO.

The design is founded on the idea"3+ 30"——a concept which proposes dividing the 80 hectare urban development into a trilogy of initial strategic project that establishes a basic structure which then serves as a solid foundation for a number of further projects, initiated in part by the municipality as well as by private investors and residents. A total of 47 subprojects with a combined total budget of 280 million Euros have since been developed. In addition to the various squares, boulevards and parks, a family of bridges are realised to improve connections between the urban and districts along the river.

Square & Park

广场和公园

The Avenida is one of the most important roads in the centre of Madrid and is characterized by its impressive environs. The motorway lies at the boundary between one of the most densely-built residential quarters and the Casa de Campo. By relocating the road in a tunnel and providing underground parking for 1,000 vehicles, it was possible to convert the space into a garden, benefitting the local residents in particular.

The design takes a journey to Portugal as its theme——the extension of the Avenida de Portugal leads towards Lisbon, in the process crossing a famous valley for its cherry blossoms. The abstraction of the cherry blossom as a design element of the park, the planting of different kinds of cherry trees to extend the period in which they flower, the reinterpretation of the Portuguese paving and the connection of the space to its surroundings have led to the creation of a popular public space.

The City Palace was built as a Baroque ensemble with a strict choreography that connected the Royal Palace with the hunting grounds and the fruit and vegetable garden at the other side of the river. Through the infrastructure changes of the fifties, the orchard was turned into a transportation hub. Contrary to the initial tendency to create a historical reconstruction, the Huerta is now a modem interpretation of the orchard. The motive of the hortus conclusus has been formed with a wide variety of fruit trees in groups, formed from skipping ranks. Fig trees, almond trees, pomegranate and more of such plantings symbolize paradise in the past. In recent decades, undertunneled river is meandering again through the room. Its source and the mouth are specially shaped.

The dominating motive for the biggest part project is the water. The park is based on different emotions and landscapes in context of the water that makes this element feelable and explorable. The system of streams is running through the park and will form different spaces and motifs in the crossings. The different streams have their own characters.

Square & Park

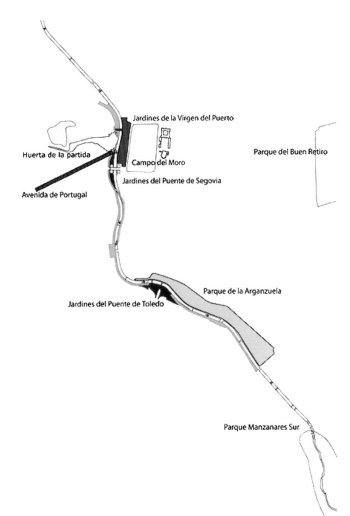

099

广场和公园

城市之丘
The City Dune

项目档案　　　　　　Project Facts

设计：SLA　　　　　　Landscape Architecture：SLA
项目地点：丹麦，哥本哈根　Location：Copenhagen, Denmark
面积：7 300 平方米　　　Site Area：7,300 m²
完成时间：2010　　　　Year：2010

数年来，哥本哈根的港区因为低质量的办公楼，落后的购物中心，糟糕的设施以及缺少公共空间而备受批评。瑞典斯德哥尔摩私人银行（简称 SEB）计划在这里利用本斯托费格德街与卡尔波德·布里吉街交汇处地下停车场上方的场地建设他们的北欧总部。SLA 接受了这项委托，旨在设计一处能将银行总部与周边环境、海港乃至哥本哈根市融为一体的城市空间。

银行大楼前方的开放空间不一定总是朴实无华，暗淡而且让人不可靠近的。所以，SLA 打破传统，将银行大楼前方的场地设计成一处服务大众和银行工作人员的绿色和友好空间。这个空间占地 7 300 平方米，宛若一个巨大的沙丘活雪堆，贯穿于建筑物之间，又围绕着建筑物，营造出空间的连贯性，并且具有较强的可持续性和可达性。此外，一片高出周围地面 7 米的绿地，为自行车和行人提供了空间。这个项目是由层叠的白色混凝土构成的，其设计灵感来源于丹麦北部沙丘和斯堪地纳维亚冬天的雪丘。层叠的造型和地形的轮廓从技术和功能层面上满足了种植所需要的排水，可达性和光照的要求，并能形成对植物根部有利的结构层。这个项目同时提供了丰富多样的步行路线，满足了 SEB 客户和员工的需求。

为了获得完整的亲身体验，你需要自己穿越这座城市之丘。当你穿越这个区域时，会发现空间在进行着不同方向的转变，从而产生新的空间关系。从本斯托费格德街一路上升，经过一段 300 米的斜坡时，空间会逐渐展现开来。当你回望城市，建筑群以框景的形式将哥本哈根呈现出来。

Square & Park

广场和公园

The harbor front of Copenhagen has through the years been widely criticized for being the site of low quality office buildings, introvert shopping malls, bad infrastructure, and few public spaces worth using. Here, above an underground car park on the most traffic-heavy corner of Copenhagen, the Swedish SEB Bank chose to erect its Scandinavian headquarters. SLA got the assignment to create an urban space that could tie the new headquarter together with the surrounding area, the harbor, and the rest of Copenhagen.

An open space in front of a bank does not necessarily need to be anonymous, grey, and void of people. On the contrary, SLA designed the area as a green and welcoming "open foyer" for the public and employees of the bank alike. The result is a sustainable and fully accessible urban space covering an area of 7,300 m^2. Like a giant dune of sand or snow it slips in between the buildings, thereby creating a spatial coherence in the design. Simultaneously, the urban space, elevated 7 meters above the surroundings, ensures the mobility of pedestrians and cyclists.

The City Dune, as the urban space, is made of white concrete, borrowing its big, folding movement from the sand dunes of Northern Denmark and the snow dunes of the Scandinavian winter. The folding movement and the contour of the terrain not only handle functional and technical demands from drainage, accessibility and lighting to plantation and the creation of a root-friendly bearing layer. It also offers a variety of routes for customers and employees of SEB as well as ordinary Copenhageners, creating an ever changing urban space. To fully experience The City Dune, one has to physically move through it. The space evolves and opens up in different directions, creating new spatial connections in the process. When ascending from Bernstorffsgade, the space gradually unfolds as you walk along the 300-meter-long and winding incline. Looking back against the city, the buildings frame a solid cut of Copenhagen.

Square & Park

比拉容马尔公园
Birrarung Marr Park

项目档案	Project Facts
设计：ASPECT 工作室	Design：ASPECT Studio
项目地点：澳大利亚，墨尔本	Location：Melbourne, Australia
完成时间：2010	Year：2010

这个项目创造了一个活跃的城市空间，满足了社区举行各种盛会和雕塑展览的需求。在平时，这个公园也为社区的人们提供了一个休闲的好去处。这个公园拥有流畅的走道和单车道，并且与城市中心有很好的衔接性。这个项目获得了2010年沃尔特伯利格里芬城市设计奖。这个公园新种植了200余株树木，而且种植大量当地的树木，这些树木的防旱性很强。

Birrarung Marr was envisaged as an active, urban space, catering for community festivals, changing sculpture exhibitions and major events such as Circus Oz and the Moomba Wateriest, while providing for passive recreation at other times. The park is a part of the Capital City Trail, providing a continuous walking and bike path along the Yarra River as well as linking the city centre to the sports precinct (MCG and Tennis Centre) in the south east. It had received the Walter Burley Griffin Award for Urban Design in 2010.

In addition to the mature elms along the river bank and at Speakers Comer, about 200 new trees were planted in the park during its construction. Unlike the water-thirsty European style gardens south of the river, Birrarung Marr was planted with hardy natives that require little watering.

Square & Park

广场和公园

五船坞广场,加菲尔德街
Five Dock Square, Garfield Street

项目档案 Project Facts

设计:ASPECT工作室,Bates Smart Design:ASPECT Studios, Bates Smart
项目地点:澳大利亚,悉尼 Location:Sydney, Australia

这个项目包括一栋垂直的建筑,这栋建筑又包括一个首层的超市,二层的公立图书馆,高层的住宅单元。景观包括一条公共道路,街边装饰,还有一个私人的向北的院落,这个院落位于公立图书馆之上,通过一系列盒子状的玻璃灯和图书馆相连接。三面加高的平台最大限度地增强了北面的光线。

栽植的落叶树木不仅为院落层的公寓提供了屏障,而且充当了天然的"太阳能空调",调节室内的温度。大量耐旱的植物点缀其中,成为院落的质地,丰富了院落的颜色。

Square & Park

This mixed-use development in Five Dock comprises a vertically stacked building program including a ground level supermarket, a public library on the first level and residential apartments on the higher levels. The landscape consists of a public lane, streetscape improvements and a private communal north facing courtyard. The courtyard sits atop the public library and is connected to it through a series of glass light boxes which are lit from within and reveal themselves as lanterns in the landscape. The courtyard is composed of a series of three raised platforms which allow maximum access to north light.

Deciduous trees provide screening and solar conditioning for the courtyard level apartments. Mass plantings of drought tolerant perennial species frame the platforms, providing texture and colour.

广场和公园

 **墨尔本会展中心**
Melbourne Convention Centre

项目档案	Project Facts
景观设计：ASPECT 工作室	Landscape Architecture：ASPECT Studio
项目地点：澳大利亚，墨尔本	Location：Melbourne, Australia
面积：6 000 平方米	Site Area：6,000 m²

新的墨尔本会展中心的设计重点在于建造一个综合性的公共空间，并且大型的建筑和设施一应俱全。这种设计保证了场地和建筑之间较高的可达性。整个项目包括会展中心、酒店、零售、住宅、公共广场、步行街、遗产景观和现有的已翻新的展览中心公园。

城市设计的运用创造了一个"六星级绿星"会展中心。另外，这个项目也注重兼容性和可持续发展性、高质量性。

这个项目设计获得了很多奖项，包括 2009 年和 2010 年 UDIA 国家环境优秀奖和 2009 年 BPN 可持续性奖。

Integration of the public realm with large scale buildings and facilities is the key to the design for the new Melbourne Convention and Exhibition Centre and the adjacent South Wharf mixed-use precinct. The design allows continuous and seemingly uninterrupted access through the site and amongst the buildings. This is a challenging task for a precinct that contains a convention centre, hotel, retail, residential, public forecourts, promenades, laneways, heritage landscapes and the now refurbished Exhibition Centre Park.

Far-reaching WSUD initiatives assisted in the overall water treatment enabled the Convention Centre to attain a 6-star Greenstar rating, the highest sustainability rating achieved in Australia for developments of this type. The design of the Melbourne Convention Exhibition Centre exemplifies our approach to collaboration, sustainability and enduring quality.

The design has received various awards including 2010 UDIA National Environmental Excellence Award, 2009 UDIA Environmental Excellence Award and 2009 BPN Sustainability Award.

Square & Park

广场和公园

Square & Park

广场和公园

海滨公园
Waterfront Park

项目档案

设计：Thomas Balsley
项目地点：美国，佛罗里达州，坦帕
面积：32 375 平方米

Project Facts

Design：Thomas Balsley
Location：Tampa, FL, USA
Site Area：32,375 m²

这个项目占地八英亩，拥有独一无二的设计和可持续发展的建造，以及很多鲜明的特点，包括污水再次利用以及LED灯。公园里有一个大草坪，通常人们可以坐在这里欣赏大型的表演。另外还有小狗休息区、洗手间、凉亭和管理处。公园的东面有一个百叶窗型的喷泉，而西面有一个雾状喷泉。在公园的游乐区域，最显眼的就是NEOS 360 Ring，它将电子游戏和有氧运动结合起来，在美国的西南部开创了一个先例。设计师Thomas Balsley说："21世纪成功的城市公园设计必须平衡创意和革新，并且要是切实可行的。对于这个项目，我们建造了各种大小的空间，从大的草坪到小的观望台和花园房，这些空间分别满足了不同的需求。"

在公园里有一个游玩区和一个小狗休息区，其设计灵感来源于艺术博物馆。草坪上的木筏，长椅，野餐桌更增添了公园的创意性，促成了公园设计的成功。大量的米草属植物增加了公园的本土气息。草坪，花园和喷泉都是由一个水系统控制的。鲜明的LED喷泉地灯和公园其他的照明灯丰富了公园的夜景以及附近居民的生活。

Square & Park

广场和公园

Square & Park

The new eight-acre Curtis Hixon Waterfront Park features a unique urban design, sustainable construction, and operational features including reclaimed water for irrigation and LED lighting. Park amenities include the Great Lawn with flexible perimeter seating to accommodate a wide range of programming and performances, a dog run, a kiosk with restrooms, and a pavilion building with restrooms, park offices, and space for a future vendor. There are also two interactive fountains: the Louver fountain is located on the east side of the park along Ashley Drive and the Mist Fountain is located on the west side the park along the Riverwalk. The park's new playground features an interactive NEOS360 Ring, which combines video games with aerobic exercise and is the first of its kind in the South East United States.

The designer Thomas Balsley says that Successful 21st century urban parks must balance creativity and innovation with proven recipes for design programs. For this project, we've created spaces ranging in scale from large open lawns to small intimate overlooks and garden rooms, able to accommodate large or small events.
Located along the river are a contemporary play area and urban dog run which take their sculptural cues from the Museum of Art. Timber lawn rafts and lounge chairs, and picnic tables with distinctive swivel loungers make up the innovative array of park furniture that is critical to the park's success and a hallmark of Balsley's design.
Masses of spartina and tree groves make up a large portion of the park's native plantings. Lawns and garden areas as well as the fountains operate on a reclaimed water system. Distinctive LED fountain pavement lights and others throughout the park extend its nighttime curb appeal and downtown activity.

广场和公园

Square & Park

广场和公园

巴塞罗那 EI Jarddin Botanico
El Jarddin Botanico de Barcelona

项目档案

设计：Bet Figueras, Artur Bossy
项目地点：巴塞罗那

Project Facts

Landscape Architecture：Bet Figueras, Artur Bossy
Location：Barcelona

应该如何规划植被呢？根据地理标准来制定整个规划图，根据五大地中海区域的植物来分类，这些都是非常重要的。景观中的植被虽然经过人工设计，但是却不能丧失在自然界中的原有姿态。

另外还有一点需要考虑的是，山体本身的地形环境也会影响到园林中不同的植物区域。设计上必须尊重原有的网络状小径，尽量避免大面积的土地改造，才能保留区域原有的地方性特色。

在原有的空间和山体滑坡的基础上，建造了一个三角形的网络结构。网络状的小径将整个空间分成了71个相互联系的小空间。

Square & Park

广场和公园

How the vegetation was to be structured? It was important to plan the layout according to geographic criteria, grouping the plants according to the world's five Mediterranean regions. Within these regional groupings, moreover, the plants should be combined according to ecological affinity, that is to say, recreating landscapes as they are found in nature.

The second consideration involved creating a project in which the mountain itself provided the topographic conditions for establishing the different plant areas in the garden. This entailed designing the network of paths around the natural relief and avoiding large earth moving operations as far as possible.

The result was a triangular-shaped network adapted to the available space and to the mountain slopes. This mesh of paths marked out the 71 spaces containing the principal plant communities found in Mediterranean climate regions all over the world.

Square & Park

广场和公园

蒙特伊克花园及凉亭
Montjuic Garden and Pavilion

项目档案	Project Facts
设计：Fondarius	Landscape Architecture：Fondarius
项目地点：西班牙，巴塞罗那	Location：Barcelona, Spain
面积：1 500 平方米	Site Area: 1,500 m²

根据市政府的规定，在城市和海边之间的蒙特伊克建造大楼是不可能的，但是市政府却热衷于将一个绿色区域融入城市中。设计师通过对现存结构的再利用和恢复，使这片绿地充满生机，纳入城市休闲网。最近的此类范例就是上世纪70年代Miramar建筑改建。设计师们将稍高的地区改建为可以远眺地中海和港口的码头，码头上还有一个小酒吧；将较低的地区改建为花园，花园位于餐厅背面与山坡挡土墙之间。选用的建材体现了地方特色，例如轻质钢和玻璃结构；平台采用的柚木，墙壁上和广场上采用板岩等。折衷、优雅的设计风格与当地其他现代的酒吧和餐厅类似。其他细节则是受其他现代建筑和蒙特伊克名胜——Mies van der Rohe's Barcelona pavilion 的启发而完成的。

Although municipal regulations make building on Montjuic——the mountain between the city and the sea——almost impossible, the Barcelona municipality is nonetheless keen to incorporate this green area into the city's leisure network. The only way to do so is by reusing and revitalizing existing structures in order to give new initiatives to a place. A recent example is the conversion of the Miramar building. It dates from the 1970s and is a well known point of reference in the city, visible from far away. The architects took into account the two completely different "faces" of the site they had to redefine. They turned the higher panoramic level into a maritime deck with a little bar overlooking the Mediterranean and the port, while the lower area is now an intimate garden (with bamboo, of course), between the rear facade of the restaurant and the rocky "retaining wall" of the hill. The ambivalence of the genius loci (high up on the mountain1 or in the shelter of the slopes') is reflected by the tectonics of the used materials: the light steel and glass construction of the bar and the teak wood of the platform relate to the sea and the sky while the slate stone of the walls and the square platforms in the garden breathe earthly solidity.

The design connects with the styling of other modern bars and restaurants in the area: eclectic and informally elegant. The detailing, though, was inspired by an older modern building and another major Montjuic attraction: Mies van der Rohe's Barcelona pavilion.

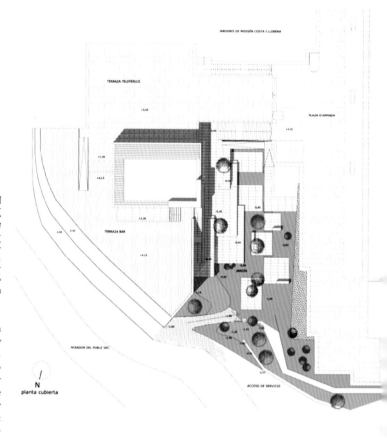

Square & Park

广场和公园

Square & Park

广场和公园

Ballast Point 公园
Ballast Point Park

项目档案

设计：McGregor + Coxall
项目地点：澳大利亚，悉尼
面积：25 000 平方米

Project Facts

Landscape Architecture：McGregor + Coxall
Location：Sydney, Australia
Site Area：25,000 m²

McGregor + Coxall 和其咨询团队赢得了为该公园进行设计发展、文案、场地监督等工作的机会，项目的设计理念即在保留场地历史的同时也为未来营造一座公园。

新建的公园具有以下特点：使用混凝土、砖石等其他建筑碎料填充成型的镂空墙体，形成逐级上升的观景平台，以获得广阔的全景视角；收集了 34 000 株乡土植物的种子进行栽种，该地区最终将会完全由绿色覆盖；对原加德士公司场地中的楼梯进行简洁的改变，为场地中的不同平面提供了极富戏剧性的连接；种植着湿地植物的水塘对进入港口前的水流进行了清洁，并为本地鸟类和青蛙提供了栖息地。

狈角公园是一座令人极为鼓舞的公园，它传递出对悉尼未来的乐观精神。在这片土地上，原本覆盖着澳大利亚污染最为严重的重工业残留下的砂石，现今却成为了对悉尼而言意义深远的新景点。

Square & Park

广场和公园

Square & Park

广场和公园

McGregor + Coxall and their consultant team were awarded the design contract for design development, documentation and site supervision for the park. The design philosophy was to acknowledge the site's past history while providing a park for the future.
The new park features: Concrete, brick and crushed building materials provide the fill to sculpted gabion walls which retain a sequence of stepped viewing terraces with sweeping panorama views. 34,000 plants grown from seeds collected locally will eventually clothe this headland in green. Elegant reinterpretations of the old Caltex site staircases provide dramatic access between the various site levels. Water ponds with wetland plants clean all the site water prior to entering the harbour and provide a habitat for local birds and frogs.
Ballast Point Park is an inspirational park conveying an optimism for the future of Sydney. For here, an area that was once blanketed with the grit of one of Australia's heaviest industrialised areas, now exists a significant new addition to the Sydney experience.

广场和公园

南森公园
Nansen Park

项目档案　　　　　　　　Project Facts

设计：Bjorbekk & Lindheim　　Design：Bjorbekk & Lindheim
项目地点：挪威，奥斯陆　　　Location：Osllo, Norway
面积：200 000 平方米　　　　Site Area：200,000 m²

在上个世纪40-60年代，这片旧有的带有各种植物的绿地成为了奥斯陆国际机场，直到1998年，机场迁出，这里便成为了一片废弃的荒地。10多年之后，根据过去旧式园林的自然形式和机场机器般的直线跑道，这里被重新改造成一个新的环境——南森公园（Nansen Park）。如今该公园给周边带来了6 000个新住宅的重建计划，并为周边15 000个居民提供了工作场地。

机场迁出后，留下这块约有1 000英亩的半岛等待改造，因此，这项工程也成为挪威最大的一项工业改造工程。工程要求在此建造一个功能、设施完整的公园和具有奥斯陆城市中心特色的新的社区。住宅和办公地点的土地卖给了私人开发商，Statsbygg（挪威公共建筑和财产管理局）和奥斯陆城市政府负责这一地点的基础设施和景观建设：对受到污染的土地，公路建筑物和技术设施进行处理，规划新的景观架构和公园结构，这其中包括建造新的娱乐区域、规划步行道路网络等。

芬兰景观设计师Helin和Siitonen赢得了该项目的设计竞赛，他们的方案：整个区域被一条环形公路包围，使该区域形成了一个碗状景观，中间是一个公园，公园周围有七条公路，像手臂一般向各个方向延伸并到海岸。在规划阶段，他们就水上设施设计和水处理与工程公司Norconsult和德国公司Atelier Dreiseitl进行了合作。2008年，公园建成并以挪威北极探险家、科学家、人道主义者、外交官、诺贝尔和平奖获得者弗里特约夫·南森（Fridtjof Nansen）命名，而他早前就住在这附近。

南森公园大约有200 000平方米，它将为市民提供一处引人入胜的活动集会、休闲的地点，它的三面都与奥斯陆海湾（Oslo Fjord）相邻。开放的园林，远处山的轮廓，会给人带来一种强烈的平和感，独立空旷的区域也将被试着融入到新的园林中。我们希望公园有着独特的个性，同时给人一种简单随意的感觉。

公园建造的新起点从旧式的机场控制塔和南面的候机楼开始，塔楼广场被建设成公园的一个重要的入口。水系也从这里由北向南延伸贯穿整个公园，水系的设计也映衬出直线景观和自然景观之间的多样性，同时也突出了水池、涓涓小溪或是小小瀑布。水域开始处是一条窄窄的沟渠，沟渠中有一个钢制框架，水通过这条沟渠向下流向一个1.5米宽的由水泥制成的水道。然后，水又从这里流入一个更大的水池，最后流入有6 000平方米的中心湖内。大部分的水都是清澈见底的，然后被泵抽回到塔楼广场。南森公园将从周围住宅区域和公路区域获得地表水，公园中也建造了许多开放的绿色洼地，这些洼地将水带入中心湖。而增加的生态沙子过滤器、机械过滤器和水泵保证了水的质量。

广场和公园

An old cultivated landscape with much variation and beauty was levelled into Oslo's international airport in the 1940-60's. In 1998, the airport moved out and left behind a depressing wasteland. After 10 years, a new environment has been created, with visual references to the old natural forms of its landscape history, and in a visual dialogue with the more recent machinelike linearity of the airport runways. The Nansen Park now awaits 6,000 new housing units and work spaces for 15,000 people along its perimeter.

The old Oslo International Airport at Fornebu left a peninsula of almost 1,000 acres to be transformed. This has resulted in the largest project for industrial reclamation in the country. It was decided that a new park should form a functional focus and identifying centrepiece of a new community 10 kilometres from downtown Oslo. Plots for housing and offices were sold off to private developers, while Statsbygg (the Norwegian Directorate of Public Construction and Property) and the City of Oslo undertook responsibility for infrastructure and landscape: the treatment of polluted grounds, the building of roads and technical infrastructure, as well as the planning of new landscape forms and the building of a new park structure. This included the establishment of new recreational areas, buffer zones along areas for nature preservation, a network of pedestrian walks, in addition to the unifying central Nansen Park itself.

Square & Park

广场和公园

Finnish architects Helin and Siitonen won a city planning competition in 1998, forming the planning premises for the whole peninsula: a ring road establishing the perimeter for a mildly bowl-shaped landscape with a centrally located park and seven green arms reaching out towards the sea in all directions. An architectural competition in 2004 was won by landscape architects Bjobekk & Lindheim. In the planning phases, the designers have cooperated with the engineering firm Norconsult and the German firm, Atelier Dreiseitl, for water design and water treatment. In 2008, the park was named after Fridtjof Nansen, a polar explorer, scientist, humanitarian, diplomat and Nobel laureate.

The central Nansen Park of approximately 200,000 square meters has been designed to serve as an attractive and active meeting place for all those who will live at and use the new Fornebu. It is bordered by the Oslo Fjord on three sides. The openness of the landscape, as well as the distant contours of the hills to the north and west of the city, gives a strong and peaceful sense of the sky, a separateness and spaciousness which we have also tried to instill in the new landscape.

The designers have attempted to combine the quiet calm of the extensive views and the harmonious forms with multiple options for activities and physical exercise. They also have wanted to give the park a strong identity, while at the same time giving it simplicity and timelessness.

Square & Park

The old airport control tower and the former terminal building to the north is the important staring point for the new park. The designers have placed the Tower Square here in order to mark the tower as an important entry. From here, a waterway stretches from north to south through the entire park. The design of the waterway itself also reflects the playful variations between straight and organic forms, still reflecting pool surfaces and streaming or falling water. It starts with a narrow water channel running over rippling coloured glass within a frame of corten steel. From here, the water runs down a 1.5-meter-wide water channel made of concrete, cast in situ, with edges and small bridges also in corten stell. The water is then led into a larger basin with a precise, hard side and a softer, green organic side with thresholds and rapids of 40-50 cm before it empties into the large Central Lake of 6,000m^2. Surplus water is led through a overflow into a infiltration area before ending in the fjord. Most of the water is cleaned and pumped back up into the Tower Square. The Nansen Park will be the recipient of surface water from the adjacent housing areas and roads. Open green swales have been built in order to carry the water down towards the new Central Lake. Biological sand filters, mechanical filters and pumps clean and air the water sufficiently to ensure good water quality.

广场和公园

HtO 城市海滩
Popular Urban Beach HtO

项目档案	Project Facts
设计：JRA 建筑事务所	Landscape Architecture：Janet Rosenberg + Associates
项目地点：加拿大，多伦多	Location：Toronto, Canada
面积：22 300 平方米	Site Area：22,300 m²

HtO 是一处沿着多伦多滨水地区的城市海滩，人们在这里可以远离市中心的骚乱，放松心情。HtO 是一个季节性变换的公共空间，深受人们喜爱。不仅如此，它同时也是城市未来海边发展的一种催化剂，具有较高的设计标准。HtO 海滩由一处废旧的工业用地翻修而成，是一个受欢迎的品牌公园，为当地社区和旅游准备，有效地通过游客和周边滨海的活力和色彩带动了周边经济。

HtO 的设计不管从视觉感受的角度还是身临其境的体验上来说，都是一个典范，因为它将城市与港口码头相联系的步行道设计得十分活泼流动，在同类型的设计中，实属首创。首先，该设计对人们的视觉感受产生了巨大的冲击力。当人们走进这个公园，起伏多变的地形引导人们要先跨越一段草坡，之后便来到沙滩、面对大海，嘈杂的城市和繁忙的高速公路就被迅速抛在了脑后；从人们身临其境的亲身感受层面上来说，公园为人们提供了充足的场地来进行身心的放松。公园更加与众不同的是，它不像其他海滩的设计那样，人们只能在有限的范围内进行步行活动，在这里人们既可以在沙滩上享受日光浴、在草坪上野餐聚会，也可以在硬质铺装的步行道上骑自行车，这样便拓宽了河滩的功能，使这片区域成为名符其实的市民休闲广场。

此外，HtO 的设计还为港口及码头的设计翻开了新的篇章。HtO 作为多伦多的第一个港口公园，它的建成，为满足已经普遍存在多伦多港口的公共开放空间中的人们的活动，提供了一个很好的借鉴模式。如今，这片滨水空间不论是在地平线上欣赏，还是从空中俯视，都可以说是多伦多的一个标志，并与整个城市和谐统一。HtO 这个名字，也成为了安大略湖沿岸滨水公共空间的品牌，主导着今后滨水公园的设计方向。

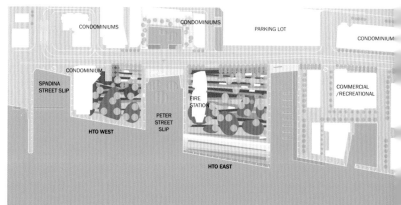

Square & Park

广场和公园

HtO is a popular urban beach along Toronto's waterfront. It was designed with the intention of attracting people to the water's edge and animating Toronto's shoreline with activity. Multiple yellow umbrellas enclosed in sand and green dunes make the space very iconic from street level and from the sky while the name, which is a play on the formula for water, HtO, is a way of branding the park.

HtO creates a more fluid passage between the downtown and the waterfront, visually and functionally. Visually, the topography of the park allows visitors to go uphill through the green berms as they enter the park and then they descend down towards the beach and the lake, thereby creating the sense that the city and the elevated expressway are left behind. Functionally, the park draws people to the waterfront by providing sufficient opportunities for shaded relaxation or leisurely activities. Unlike anywhere else within walking distance, visitors to the park can spend the day sun tanning, picnicking or bike riding in a space that makes unique use of hard and soft surfaces.

HtO has also helped to establish a new language along the waterfront. It is the first park to be completed in the area for some years and it sets the tone for the development of public open spaces that is currently happening at Toronto's waterfront. The urban beach becomes very iconic both from street level and from aerial views, giving a new identity to the community. The name HtO itself has been devised as a branding tactic that helps to position the park as an important addition in the sequence of public spaces along Lake Ontario.

Square & Park

广场和公园

糖果沙滩
Sugar Beach

项目档案	Project Facts
设计：Claude Cormier 建筑事务所	Architect：Claude Cormier Associates
项目地点：加拿大，安大略省多伦多	Location：Toronto, Ontario, Canda
面积：8 500 平方米	Site Area：8,500 m²

糖果沙滩是多伦多第二大城市海滩。这个海滩设计跟随城市沙滩设计的新潮流，其创作灵感源自棒棒糖。在贾维斯湾的楔形沙滩上布置了几十把糖果颜色的遮阳伞，仿似染上了西风中飘扬的红路糖果飞沫。设计与未来的滨水散步道和广场（将用于举办常规和临时活动）融为一体，吸收了多伦多新兴的景观特性中的部分永久元素，如沙滩、岩床、树木、水、城市地平线和城市工业历史的痕迹等。

Sugar Beach is the second urban beach proposed for Toronto's downtown waterfront, and the latest addition to the amber necklace of Toronto's lakefront beachscape. It is a sequel to HtO, the waterfront's first beach park. The proposal for Jarvis Slip playfully recomposes other signature elements of the city, with Toronto playing the role as its own design precedent. Tinted by sugar spray carried on westerly breezes from the neighboring Redpath Sugar Factory, a series of hard rock candies with colored stripes and dozens of pink umbrellas are scattered across a sandy wedge of beach along the Jarvis Slip. Integrating the future Waterfront Promenade, along with a plaza for programmed and unprogrammed events, the design playfully adopts some of the most enduring elements from Toronto's emerging landscape identity——beaches, bedrock, trees, and waters as well as the urban horizon and a trace of the city's past industrial mood.

Square & Park

广场和公园

Square & Park

广场和公园

加拿大文明博物馆广场
Canadian Museum of Civilizations Plaza

项目档案	Project Facts
设计：Douglas Cardinal	Design：Douglas Cardinal
面积：2 900 平方米	Site Area：2,900 m²
完成时间：2010	Year：2010

加拿大文明博物馆包括两个展馆，展馆的建筑风格很能突显当地的地形特色。公共展馆重现了冰川的奇妙风貌，而另一个展馆则象征着宏伟的加拿大地盾。广场的布局和大小很好地协调了周围的博物馆建筑和与之隔江相望的加拿大的国会大厦。为了改善广场的游客量和受欢迎度，设计师延伸了博物馆原始的设计概念，通过广场的草的变化来诠释加拿大从荒芜发展至今的历史。

绵延起伏的山丘，蜿蜒的小径，不同形状的草地，与博物馆柔软的曲线轮廓相得益彰。这样的设计，不仅增加了当地的生物多样性，减少了热岛效应，而且提供了新鲜的空气，丰富了游客们的游玩体验。

The Canadian Museum of Civilization is comprised of two pavilions, a startling embodiment of the country's distinguishing geographical features. The public display wing replicates the dramatic effect of the glaciers; the contours of the curatorial wing symbolize the majestic Canadian Shield; and the open plaza simulates the vast Great Plains. The layout and sheer size of the plaza were planned in such a way as to visually incorporate the Museum buildings and the Parliament Buildings perched across the Ottawa River. However, the Plaza's lack of appeal had left it empty of visitors for much of the year. To remedy the situation, the designer extended the Museum's original conceptual metaphor, bringing to life what had long remained latent.

The prairie topography has been recreated through hillocks interspersed across the plaza and wound through by snaking pathways, suggesting both of the Museum's curves and the soft and undulating prairie landscape. The granite paving stones hug and accentuate the gently rising mounds. The introduction of an urban prairie to the environment creates a microclimate, increases biodiversity, lessens heat island effect, improves air quality, and lures the public through an appeal to the senses, without in any way hindering the stunning view that defines the site.

Square & Park

广场和公园

达尔哥诺·玛公园
Park Diagonal Mar

项目档案	Project Facts
设计：EDAW	Landscape Architecture：EDAW
项目地点：西班牙，巴塞罗那	Location：Barcelona, Spain

达尔哥诺·玛公园是一处集商业区和住宅于一体的综合性城市景观，该公园的建造是巴塞罗那城市海滨复兴的一部分。公园为市民和游客提供了通往海边的通道。公园里有游乐场地、湖泊、瀑布、露天咖啡馆、喷泉和观景护堤等，所有景观都由通向沙滩的小路连接。公园整体效果极佳，周围高层建筑里的住户从高处也能俯瞰整个公园的美景。

生态环境在整个公园的设计和建造过程中起着重要的作用。可渗水人行道将降水径流效果降到最低。当地植被的种植减少了灌溉和杀虫剂的使用频率。公园里还有地域性澄清池，海岸线周围有水生植物，可以过滤降水，这里还是当地水生动植物和鸟类的栖息地。

Square & Park

广场和公园

Square & Park

Diagonal Mar, a retail and residential complex centered on a public park, completes Barcelona's urban beachfront revitalization. Providing a gateway to the sea for locals and visitors, Park Diagonal Mar features play areas, lakes, a waterfall, an outdoor cafe, a fountain and viewingmounds, which are linked by paths that lead to the beach. The park is conceived as an abstract tapestry in plan view, and is enjoyed from above by high-rise residents.
Ecology plays a meaningful role in the park's design. Porous pavements minimize stormwater runoff. The use of native plant materials allows for reduced irrigation and pesticide use. A regional retention pond and sections of shoreline edged with aqustic plants provide first flush cleansing of stormwater, as well an a habitat for indigenous marine and bird species.

广场和公园

高线公园 2 期
Section 2 of the High Line

项目档案

设计：James Corner Field Operations, Diller Scofidio + Renfro
项目地点：纽约
面积：10 927 平方米

Project Facts

Landscape architects：James Corner Field Operations，Diller Scofidio + Renfro
Location：New York
Site Area：10,927 m²

高线公园 2 期从西 20 街到西 30 街，横跨 10 个街区，鲜明特点为宽度较窄（通常是在 9 米左右）和直线线性关系（10 个线性块）。设计在各方面与一期保持了相同的基本元素（铺装、种植、家具、照明、交接处理等），同时强调通过一系列有特色的序列空间，营造出丰富的体验观感。

野花花坛
位于西 26 大道和西 29 大道之间的野花花坛由硬质铺地、耐旱性的草坪和野花构成，花坛中种植了种类繁多的花卉以确保在生长季能一直有花开放。简洁的直线形道路沿着两旁的野花延伸开来，繁华盛开的花坛点缀在旧铁路道上，行人可以尽情地欣赏这道绿色风景线并穿梭在城市之中。

切尔西灌木丛
当参观者从切尔西草场向北移动，一片茂密的花丛和低矮的小树林便映入眼帘，它们是高线公园另一个景观的起始点。位于西 20 大道和 22 大道之间的切尔西灌木丛种植了多种多样的植物，例如美洲冬青，紫荆和其它美国本土的常绿植物，它们能够一年四季不间断为公园提供丰富的色彩变化。而另外一些较为低矮的植物，例如苔草和耐荫性灌木则强调了从草地向灌木丛的转换。

第 26 大道观景台
悬吊在高线公园东侧原有铁路线上方的观景台位于西 26 大道，这个设计旨在让人们回想起原来布置在公园上的户外广告牌。高大的灌木丛和树丛位于观景台侧面，同时参观者还可以在木质的观景平台上坐下休息，欣赏第 10 大道和切尔西的风景。

第 23 大道草坪和阶梯座椅
在西 22 大道和西 23 大道之间的高线公园拥有着更宽阔的面积，这里原来曾是铁路线的备用站点，用来装卸旁边仓库的货物。这个额外的空间被设计成一个聚会场所，并安置了一组阶梯座椅，这些座椅是用回收来的柚木制成的，并锚固在一片面积达 455 平方米的草坪上。草坪的最北端向上"卷起"，把台阶座椅上的参观者抬升到几米高的空中，以享受布鲁克林以东和哈德逊河的美丽风光。

philip a. 和 lisa maria falcone 立交桥
在西 25 大道和西 26 大道之间，建筑师在两座相邻建筑之间创造了一个微型森林，这里种植了茂密的灌木丛和小树。现在，设计师又在原有的环境上架起了一座高 2.5 米的步行道，这种处理让地面上的植物可以自由生长并且保持了原本有机起伏的地形，参观者则在茂密的树廊下穿行。

Square & Park

广场和公园

广场和公园

Square & Park

The Section 2 of the High Line runs from West 20th Street to West 30th Street (ten blocks). One of the distinctive characteristics of this area is its narrow width (typically 30 feet across), and the straight linearity (ten linear blocks). While the simplicity in design materials and the consistency with the rest of High Line design are maintained through the employment of the basic design elements (planking, planting, furnishing, lighting, access points elements etc.) prototyped in Section 1, the uniquely narrow and linear site conditions are emphasized by a series of distinctive spaces, which provide a cohesive but distinct sequence of experiences.

Adial Bench
At west 29th street, the high line begins a long, gentle curve towards the hudson river, signifying a transition to the west side rail yards. The high line's pathway echoes the curve, and a long bank of wooden benches sweep westward along the edge of the pathway, planting beds behind and in front of the benches line the curve with greenery.

Chelsea Thicket
As visitors move north from the chelsea grasslands' prairie-like landscape, a dense planting of flowering shrubs and small trees indicates the beginning of a new section of the park. Between west 20th and west 22nd streets, in the chelsea thicket, species like winterberry, redbud, and large american hollies provide year-round textural and color variation, an under-planting of low grasses, sedges, and shade-tolerant perennials further emphasizes the transition from grassland to thicket.

26th Street Viewing Spur
Hovering above the historic rail on the east side of the high line at west 26th street, the viewing spur's frame is meant to recall the billboards that were once attached to the high line. The frame enhances, rather than blocks, views of the city. Tall shrubs and trees flank the viewing spur's frame, while a platform with wood benches invites visitors to sit and enjoy views of 10th avenue and chelsea.

23rd Street Lawn And Seating Steps
The high line opens to a wider area between west 22nd and west 23rd streets, where an extra pair of rail tracks once served the loading docks of adjacent warehouses. The extra width in this area was used to create a gathering space, with seating steps made of reclaimed teak anchoring the southern end of a 4,900-square-foot lawn. At its northern end, the lawn "peels up", lifting visitors several feet into the air and offering views of brooklyn to the east and the hudson river and new jersey to the west.

Philip a. and Lisa Maria Falcone Flyover
Between west 25th and west 26th streets, adjacent buildings create a microclimate that once cultivated a dense grove of tall shrubs and trees, now, a metal walkway rises eight feet above the high line, allowing groundcover plants to blanket the undulating terrain below, and carrying visitors upward into a canopy of sumac and magnolia trees. At various points, overlooks branch off the walkway, creating opportunities to pause and enjoy views of the plantings below and the city beyond.

广场和公园

Maister 将军纪念公园
General Maister Memorial Park

项目档案

设计：Bruto (Matej Kucina、Tanja Maljevac)
项目地点：斯洛文尼亚
面积：1 500 平方米

Project Facts

Landscape Architecture：Bruto (Matej Kucina, Tanja Maljevac)
Location： Ljubno ob Savinji, Slovenia
Site Area: 1,500 m²

这座陵园是为 Maister 将军以及北部边境的战士们建造的。陵园的设计是一个抽象的三维空间，在空间中，道路呈几何状分布在剪切整齐的草地中央。这是一种对北部边境地区的象征，在那个地区，Maister 的战士们参与了战争。

紧挨着坡峰的是一面支撑面，将道路截断。支撑墙是纪念物的一部分，因为在支撑墙上有一根金属杆，上面写着在北部边界战争中为国捐躯的烈士们的名字。这里有一具真人大小的骑马人的铜像，是该陵园的整体象征。

The park dedicated to General Maister and the soldiers for the northern border was planned as an abstract three-dimensional space, where the paths lead around geometrically cut grass crests. It is an abstract representation of the crest of the northern border, for which Maister's soldiers fought.

The crest next to the road is truncated and ends with a supporting wall, which serves as a part of the memorial symbol, and there is a line of metal sticks, bearing the names of the soldiers for the northern border. The symbolic composition ends with an abstract life-size bronze figure of a horseman.

Square & Park

广场和公园

布里克斯顿广场
Brixton Square

项目档案

设计：GROSS. MAX.
项目地点：伦敦，布里克斯顿

Project Facts

Landscape Architecture：GROSS. MAX.
Location：Brixton, London

伦敦布里克斯顿广场是由三个独立的公共空间组成：Tate花园，Windrush广场和St Mathews和平花园。这些空间由一系列公共建筑组成，包括市政大厅、St Mathews教堂、Raleigh大厅和Tate图书馆，但同时这些空间又通过一系列道路被分离开来。这个项目的目的在于创造一个公众需要的高质量公共空间，为提升当地形象作出了巨大的贡献。

这个广场为伦敦这个动态的多文化的城市提供了一个充满活力的空间。能够彰显宗教，政治和知识的传统的柱子增强了这个广场的历史韵味。空间的连贯性，形象的标志，让这个广场成为举办各种活动和盛世的首选之地。

花园和广场的结合成为一种新的城市景观发展趋向。人们在这样的环境之中，将会得到更为丰富或者说矛盾的体验，既可独处，也可结伴；既是封闭的，也是开放的；既有内导性，又有外延性。生物和文化的多样性也在这里体现无遗。

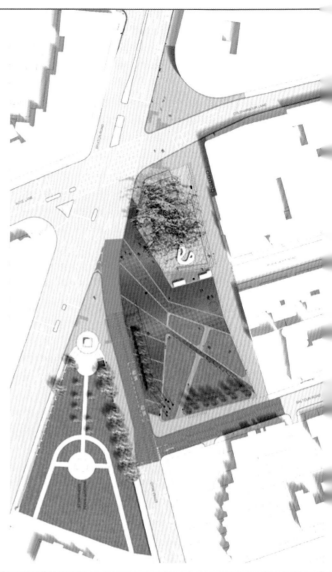

Square & Park

161

广场和公园

Brixton Square is formed by three separate public spaces: Tate Gardens, Windrush Square and St Mathews Peace Garden. The spaces are loosely defined by a series of civic buildings, including the town hall, St Mathews Church, Raleigh Hall and the Tate Library. However, these spaces are disconnected from each other and the civic buildings by a series of roads, two of which, Brixton Hill and Effra Road, carry heavy traffic. The project for Brixton Square aims to create a much needed and high quality public space that does justice to the significance of Brixton.

Brixton Square reflects on the role of civic space at the start of the 21st century as a vibrant stage for a dynamic multicultural society. Traditional pillars of society such as church (religion), town hall (politics) and library (knowledge) which are all represented within Brixton Square are no longer the priority of such a civic space. The site was lost in space and time. It needed a new agenda in terms of programme and event, and a new setting for those activities, which provides both spatial coherence and an iconic landmark.

The combination of garden and square can be reconciled into a new urban typology, which provides for opposite experiences such as solitude and gathering, enclosure and opening, the introvert and the extrovert. An integrated park and square is of both biological and cultural diversity.

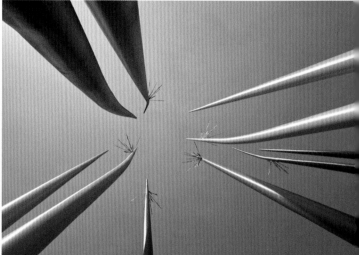

Square & Park

广场和公园

铁锚公园
Anchor Park

项目档案　　　　　Project Facts

设计：SLA　　　　　Landscape Architecture：SLA
项目地点：瑞典，马尔默　　Location：Vastre Hamnen, Malmo Sweden
面积：29 000 平方米　　Site Area：29,000 m²

铁锚公园建立在马尔默市繁茂的生活园区，它的功能是形成一个开放的、吸引人的城市空间，希望能够被每一个市民喜欢。公园是围绕一些老工业厂房建立的，并且跟它们有着相同的名字。这个公园随着季节不同，给那些建筑带来不一样的风景和用途。在瑞典自然环境多样性的影响下，在公园里建立了不同类型的生态群落，比如沼泽、桤木、栎树林和水群落等。空间的相互融合适合人们游玩和深思。

公园的建立为这个区域注入了有变化的时间感。每一天都可以感受到那些细微的变化，这使人们有了新的体验，也让他们更加体会到生活在这里的存在感与责任感。运河沿岸的混凝土基刻有浮雕。在这里，每年有113天的降雨量汇集在凹痕中，从浮雕中的雨水量和水面的反射，此地自然环境状况的改变清晰可见。

周边由1公里的现浇混凝土构成，曲折行进中创造出更加丰富的空间与时间感。四周有瑞典花岗岩石和可供小坐的树桩。交互式布局的浅草层形成了包含多个小草丛的整体，每个小草丛都有自己的形式，没有哪两个是完全相同的。

Square & Park

广场和公园

Located in the new quarter Vastra Hamnen in Malmo, Ankar Park, a luxuriant living park, has been created as an open, attractive city space for everyone. The quarter has been constructed around the old industrial harbour with the same name. The park gives the buildings sensory delights and changes with the seasons and uses. A new type of city park has been created——a hydroglyph park. Inspired by the diversity of Swedish nature, different types of biotopes have been established such as alder marshes, oak woods and salt-water biotopes with crayfish-spaces that merge into each other and spaces that invite play and contemplation.

The park invests the area with a different sense of time. Subtle changes can be registered every day, which allow for new experiences with sharpened attention and an increased sense of presence. Reliefs are impressed into the concrete along the canal. Here the rainwater is collected that falls in these parts 113 days a year. The changing nature of the area is manifested by the amount of rainwater and mirroring in the many pools on the perimeter.

Square & Park

The perimeter consists of 1km in situ cast concrete, that with its winding course creates extra space and increases time spent there. Swedish granite boulders and conical stumps for seating are placed on the perimeter, and inter-changing layers of grasses form a composition of varied patches. Each patch has its own form and is different from others.

广场和公园

奥斯陆歌剧院的屋顶
Roof of Oslo Opera

项目档案	Project Facts
设计：Snohetta	Design：Snohetta
项目地点：挪威，奥斯陆	Location：Oslo, Norway
面积：38 500 平方米	Site Area：38,500 m²

奥斯陆歌剧院和人的互动感很强，建筑的一脚直接伸进水面，屋顶与此相连并且斜着向上，给人们提供了一个活动空间。屋顶主要采用大理石和花岗岩，像一座出水的冰川，看起来异常惊艳美丽。游客可以在屋顶上漫步，饱览奥斯陆的市容美景。

From the ground, the roof of the Oslo Opera House slopes steeply up, creating an expansive walkway past the high glass windows of the interior foyer. Made of marbel and granite, the roof looks beautiful, like a glacier. Visitors can stroll up the incline and stand directly over the main theater and enjoy views of Oslo and the fjord.

Square & Park

广场和公园

Square & Park

广场和公园

索伯格塔及休闲区
Solberg Tower & Rest Area

项目档案　　　　　　　　　Project Facts

设计：Saunders Architecture　　Architect：Saunders Architecture
项目地点：挪威　　　　　　　Location：Norway
面积：2 000 平方米　　　　　Site Area：2,000 m²
完成时间：2010　　　　　　　Year：2010

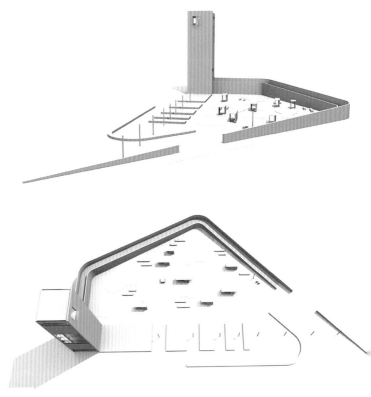

索伯格塔及休闲区位于挪威的萨尔普斯堡。这里具有南挪威平静的绿色平原风光，常有当地游客和瑞典游客到访。业主对项目没有太多的限制，因此建筑师可以非常自由地创作。设计中将使用者放在第一位，人们可以在这里停留并欣赏附近森林和海岸线的绮丽风光，避开临近高速公路的噪音，享受一个绿色休息空间。

矮墙划分出2 000平方米的用地，春季鲜花点缀庭院，庭院里还有7个凉亭，和那道矮墙一并展示当地的艺术和石刻。在观光塔的周围还有很多不为人知的石刻作品，通常大家只想着到下一个目的地而忽略了它们的存在。这里还能作为艺术家的临时展览场地。人们可以通过电梯或者楼梯上到9层，共30米高的塔顶，在高处观赏平原之美。塔叫索伯格，意思是太阳山，期望人们登顶后可以看见奥斯陆峡湾动人心魄的景色。

在设计上，考虑到周围的景色，遵循最少的原则，借鉴当地建筑，主材是用锈蚀钢板做成的墙面和用木材铺就的平台，板岩和细沙石组成了地面。当地的自然和历史通过建筑得以强调，对各方需求产生了直接和多元的回应。

Square & Park

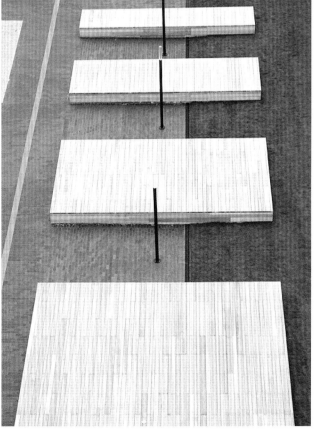

广场和公园

Square & Park

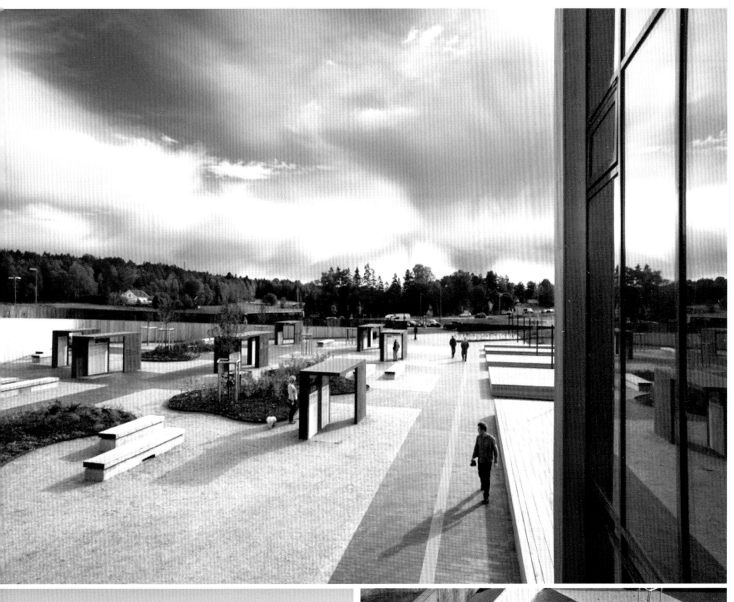

The flatness of the landscape meant that the beauty of the surrounding nature could only be enjoyed from a certain height, so the creation of a tower quickly became a main part of the brief. The ramp's asymmetrical walls rise from 0-4m, then form a 30m simple nine-storey-tall structure on the site's northern edge, including only a staircase and an elevator. Named Solberg, which translates into "Sun Mountain", the towers aerial views towards the nearby coastline and the Oslo fjord are truly dramatic. Finally, the design's style and aesthetic was developed in relation to the environs' existing architecture; minimal and geometrical contemporary shapes were chosen, contrasting the local farming villages' more traditional forms. The main materials used were beautifully-ageing CorTen steel for the exterior walls and warm oiled hard wood for the courtyard's design elements and information points. Local slate and fine gravel pave the ground level. Underlining the area's natural and historical attractions, supported by strong architectural forms, Saunders produced a complex, in direct response to both the clients and site's requirements.

广场和公园

埃斯比约海滨长廊
Esbjerg Beach Promenade

项目档案

设计：Nathan Romero, Sofie Willems
项目地点：丹麦，埃斯比约

Project Facts

Design：Nathan Romero, Sofie Willems
Location：Esbjerg, Denmark

这个新的海滩长廊不管是在内容上还是功能上都相当有创意，对当地的居民和游客都具有吸引力。这里的环境氛围特别好，就像是一个磁铁，有足够多的游玩，运动和吃冰激凌的空间。不管在什么样的天气下，只要人们来到这里，都会有独一无二的体验。
长廊总长为700米，末端是为海员和冲浪的人新建的俱乐部。一个木制的甲板和海滩线平行而建。原来用来保护公路和房子的木制墙体被转变成一排长椅。
长廊沿线设计了很多漂亮的驻足点，为居民和游客提供了各种各样丰富的体验经历。一个木制的甲板漂浮在海里，甲板上有很多木桩。这些木桩随着风，随着海浪，成为一道亮丽的动态风景，而且能反映出浪潮的高低。部分木桩上是有抓柄的，方便人们爬上甲板。海滩上也有一些木桩，部分木桩之间有吊床。
海滩长廊与海水，海风和天气相协调，给了游客新的不同的亲身体验。在长廊末端的俱乐部为水上运动和冬天沐浴提供了一个很好的地方。主要包括娱乐区，漂浮屋和一个公共的桑拿房，在桑拿房里可以俯瞰海平面的美景。

The beach promenade in Esbjerg spans 700m along the coast and ends at the new clubhouse for sailors and surfers. A wooden deck runs parallel to the coast from where ramps and tribunes connect down to the beach. The original wooden wall protecting the road and houses from the fierce forces of the North Sea was transformed into a long bench flanking it. Along its course, the promenade offers beautiful stops and various new experiences on the beach and in water. A sea-pool, dusters of large wooden poles and a floating platform invites visitors down to the beach for a new encounter. The wooden poles appear as fixpoints in the changing landscape. Wearing by wind, salty water, flying sand and ice will mark, polish and characterize the wooden poles over time. They also stand as sculptural measuring instruments for high and low tides. Some of the poles have grips, encouraging visitors to climb them. Hammocks between some of the poles bring in a human scale and sense of belonging on the beach. The wooden platform lies as a large piece of maritime furniture on the beach at low tide, and becomes a floating island at high tide and a new sea-pool offers a different and tranquil bathing experience in the North Sea.
By letting the architecture play with water, wind and weather, visitors can experience the beautiful site in new and different ways. The new clubhouse at the end of the promenade provides a basis for water sports, sailing, surfing and winter bathing. The building contains recreational areas, changing rooms, a large workshop and a public sauna overlooking the horizon.

Square & Park

广场和公园

Ubuda 城市中心
Ubuda City Centre

项目档案

设计：Garden Studio Ltd.
项目地点：匈牙利，布达佩斯
面积：20 000 平方米

Project Facts

Landscape Architecture：Garten Studio Ltd.
Location：Budapest, Hungary
Site Area：20,000 m^2

Ubuda 是布达佩斯人口最为密集的区之一，经历了很多巨大的变化。新地铁正在建设之中，新的城市中心建设计划还包括一个交易大厅的翻新，单车道的改善以及地下停车场的建造。由 ING-hat 负责的项目包括一个高端的购物中心、一个娱乐中心、一个办公大楼、还有住宅建筑。在政府的大力帮助下，环境也得到了明显的改善。一条长达 100 米的公共道路得以建立，部分西边和东边的街道变成了步行街，以前的停车场变成了一个广场和公园。

Ubuda is one of the most densely populated districts of Budapest, and it is going through significant changes. Construction of the stations of the new metro line is in progress, and the plans include the refurbishment of Vasarcsamok (a market hall), the development of a cycle track network and an underground car-park. The development carried out by ING-hat constructed a high-level shopping mall and entertainment centre (ALLEE), an office building and residential buildings. Thank to the successful cooperation with the local government, the environment has also been refreshed. A 100-meter-long public road tunnel has been built, the delimiting eastern and western streets become pedestrian zones, and, in the place of the former parking lots, an urban piazza and park awaits citizens.

Square & Park

广场和公园

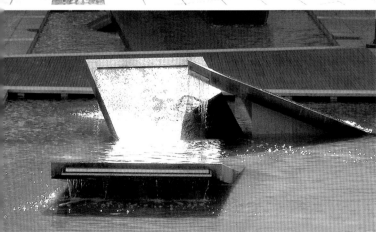

广场和公园

Dania 公园
Dania Park

项目档案

设计：Thorbjorn Andersson
项目地点：瑞典，马尔摩

Project Facts

Landscape Architecture：Thorbjorn Andersson
Location：Malmo, Sweden

厄勒海峡将丹麦和瑞典分隔开来，长长的海岸，无边的天空，构成了绝美的风景。Dania 公园就是位于这个美丽的地方。虽为海边公园，但为了方便各方向的人们进入，入口是非常多的。这里也是一个体验各种天气的好场所，游客不仅可以在这里体验夏天的炎热，秋天暴风雨的威力，还可以体验冬天的安静。

Oresund, the sound that separates Denmark from Sweden, features magnificent views of the long horizon and a wide sky. This is dramatic coastal setting for the Dania Park. As a shoreline park, it offers different ways to be close to the sea that vary from simple access to purposeful challenge. It is also a good place to experience the tempers of the climate, with Malmos sunny summer days, violent autumn storms, and calm, frosty winters with views over grey waters.

Square & Park

广场和公园

Square & Park

广场和公园

Marina 公园
Marina Park

项目档案

设计：AAAID
项目地点：西班牙，巴塞罗那

Project Facts

Landscape Architecture：AAAID
Location：Barcelona, Spain

Square & Park

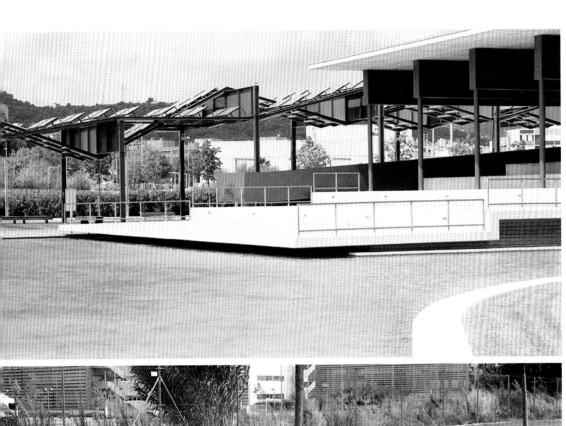

广场和公园

Square & Park

广场和公园

Square & Park

广场和公园

Nou Barris 中央公园
Parc Central de Nou Barris

项目档案

设计：A. Arriola，C. Fiol
面积：170 000 平方米
项目地点：西班牙，巴塞罗那

Project Facts

Landscape Architecture: A. Arriola, C.Fiol
Site Area: 170,000 m²
Location: Bacelona, Spain

Nou Barris 中央公园的建设是巴塞罗那东部边缘地区复兴工程的中心部分，公园的存在可以帮助曾经凄凉的东部边缘地区焕发光彩，走向繁荣。Ronda de Dalt 环路的建成加强了此处与周围地区的联系，中央公园极大地开发了该地区的潜能。

Nou Barris 区是巴塞罗那的一部分，但是却没有自己鲜明的特点，社区之间历史环境各不相同，联系性也很缺乏。公园的建立大大地增加了该区域的可发展性，缩短了区与区之间的距离。

Square & Park

广场和公园

The construction of Nou Barris Park is the central intervention in the complete renewal of Barcelona's eastern periphery. With the new international Forum organized by Barcelona on the maritime, the decision to finish the park is quite an achievement.

The Nou Barris district is a part of Barcelona with no identity of its own. With vast conglomeration of different neighbourhoods and different urban histories and weak connectivity, it extends over eight square kilometers. Its potential is greatly increased, for the well-to-do areas in the west and the District of Nou Barris around Karl Marx Plaza are now only ten minutes apart.

广场和公园

Arena 大道和阿姆斯特丹 Poort 商业街
Arena Boulevard / Amsterdamse Poort

项目档案

设计：Karres en Brands
项目地点：荷兰，阿姆斯特丹
面积：42 000 平方米

Project Facts

Landscape Architecture：Karres en Brands
Location：Amsterdam, The Netherlands
Site Area: 42,000 m²

Arena 大道和阿姆斯特丹 Poort 商业街被发展成阿姆斯特丹的第二个夜生活区域。原先的 Poort 商业街很小，而且很喧闹；而 Arena 大道虽空间很大，但是人流量却很少，建筑的入口虽少，但是内部的功能却很齐全。Arena 大道设计的重点就是打破原有的线型结构，创造一个可以容纳很多人的空间。
从北面到南面的横向连接通过在建筑之间创造空间而得到加强。公共空间的设计以自然的方式展开。天然的石头和木头做成的长椅能够给来往的人们提供一个休息的地方。这些长椅不是直线型的，在不同的方向会有轻微的曲线形美。这些区域有些是用于运动的，有些区域加以提高，种植上树木和草，也可以用作休息的地方。还有一个公共空间方便人们长时间在这里休息。
灯光设计成三维的，在大道的上空构成了一片星空。在一个主要的位置上，灯光变得更加密集，更加明亮。

Square & Park

The Arena Boulevard and the Amsterdamse Poort will together be developed into Amsterdam's second nightlife district. The current central area has two different faces: the busy, small-scale Amsterdamse Poort shopping centre and the spacious but often empty Arena Boulevard. The buildings, in which major functions are accommodated behind a small number of entrances, accentuate the impersonal character of the area. The emphasis in the design for the Arena Boulevard is on breaking up its linear character, and creating a space that is pleasant for a group of ten people, but also for a crowd of fifty thousand.

Transverse connections between the north and south side are enhanced by creating space between the buildings on the boulevard. In the design, the public space organizes itself in a natural way. Long benches of natural stone and wood mark the transition between places for movement and places to pause. Alongside these benches, the ground level rises up or falls away, giving a slightly curved paved surface. Some points are laid out for sports use, others are raised up with trees and grass creating places to relax. A public space is created that invites a longer stay and offers opportunities for various kinds of use.

A three-dimensional lighting web of cables and spotlights creates a dramatic starry sky above the Arena Boulevard. The web becomes denser and the light brighter.

广场和公园

Square & Park

广场和公园

罗马采石场设计
Roman Quarry Redesign

项目档案

设计：AllesWirdGut 建筑事务所
项目地点：奥地利
面积：5 580 平方米

Project Facts

Design：AllesWirdGut Architektur
Location: Austria
Site Area：5,580 m²

整个工程的设计灵感来源于采石场的采矿技术，该工程从顶端到底端覆盖了采石场的大面积区域。设计师设计了一系列大型的楼梯和桥梁结构，它们从采石场的顶端一直向下蜿蜒到主要的欢庆场地之上。道路两侧栏杆上生了锈的金属栏杆成一种"Z"字形状，创造出一种独特的海拔高度。略微倾斜的支柱在崎岖的地带将道路支撑起来，道路中的铁锈的颜色与采石场中浅色的石块形成对比。

The design of the whole project is inspired from the quarry mining technology. The project covers large areas of the quarry. A large set of stairs and bridge structures from the top of the quarry have been winding down to the main celebration venue. Rusty metal railings on both sides of the road in a "Z" shape create a unique altitude. Slightly-tilted pillars in the rugged terrain prop up the road. There is a strong contrast between the color of rust in the road and the light color of quarry stones.

Square & Park

广场和公园

Square & Park

广场和公园

 Maddern 广场
Maddern Square

项目档案 　　　　Project Facts

项目地点：墨尔本　　Location: Footscray, Melbourne

Maddern 广场是富士贵区一系列公共空间改善竞赛中的获奖作品。项目旨在创建一个具有催化整个区域的公共空间发展的典范。所以，在设计和建造过程中，需要详细周全的全方位考虑。墙体，座位，铺设以及交通等材料和细节的选择都要根据经济，耐用性，强度和灵活性来决定。这个广场提供了三个完全不同的空间，这三个空间既可以被分离开使用，也可以综合起来方便大型的盛会。西南方的一个广场为交易和聚会提供了一个绝佳的场所。在广场的边缘有四季常青的林木，为活动和休息提供了荫凉和新鲜的空气。墙体以及林木周围的护栏采用混凝土，增加了整个广场的层次感。沿着广场的周边还有一个木制的稍微加高的舞台，在这里可以举行小型的表演，也可以供游客暂时小憩。一个混凝土斜坡通向加高的草坪区，人们也可以通过这里进入广场内部。

Maddern Square is the first completed project in a series of public space upgrades proposed for central Footscray. Because of the exemplary and catalytic role the development needed to play, it is important that both design and construction are carefully considered. The material and detail choices for the retaining walls, seating, pavement and traffic management devices are informed by the need for economy, durability, strength and flexibility. Maddern Square provides three distinct spaces, which can be used independently or together for larger events. A pedestrian plaza on the south-west corner creates an open area for markets and gatherings. New evergreen trees provide shade for seating and activity to the edge of the plaza. Concrete retaining walls and kerbs allow levels to be maintained around the trees whilst providing definition between the key spaces of the square. A timber topped wall along the plaza edge provides a stage platform for performances, and informal seating. The concrete ramp provides access to the raised grass and gravel areas and a pedestrian connection across the plaza, as well as space for play between the upper and lower level spaces.

Square & Park

广场和公园

M 中心
M Central

项目档案

设计：360 景观设计建筑事务所
项目地点：澳大利亚
面积：3 000 平方米

Project Facts

Design：360 Landscape Architects
Location：Australia
Site Area：3,000 m²

M 中心是一个住宅大楼屋顶的延伸发展，位于悉尼 CBD 中心，独一无二的空间为创造一个充满活力的娱乐空间提供了坚实的基础。400 户的居民每天差不多有 100 户来到这里。盖板小径蜿蜒到草坪里，通向各种娱乐区域，包括一个用回收的混凝土和赤土色薄膜构成的绘画区，一个经过装饰的水池和水幕墙，秘密花园，儿童垒球区以及烧烤区域等。

M Central is a rooftop "Parkland" landscape for the residential redevelopment of the Goldsborough Mort Woolstores, Pyrmont. Unique in scale in Sydney's CBD, the rooftop is a vibrant communal recreation space and a benchmark for environmentally and socially sustainable rooftops. Up to a quarter of M Central's 400 residents use the landscape every day. The scheme features decking paths that float through expansive native grasslands leading to a variety of recreation opportunities including a painted landscape of recycled concrete and terracotta mulches, an ornamental pond and water wall, secret lawns, children's softfall play area, BBQ facilities and a sculptural arbor, cloaked in flowering climbers.

Square & Park

广场和公园

Promenade Samuel-De Champlain 长廊
Promenade Samuel-de Champlain

项目档案

设计：Consortium Daoust Lestage & Williams Asselin Ackaoui
项目地点：加拿大，魁北克

Project Facts

Design：Consortium Daoust Lestage & Williams Asselin Ackaoui
Location：Quebec, Canada

位于沉默的圣劳伦斯河和喧嚣的 Boulevard Champlain 之间，这处狭长的河岸长廊为人们提供了赏心悦目的亲身体验，不仅有无边的河岸之美，而且还有充满魅力的地形之美。项目设计首先要考虑到的问题就是废掉原有的横跨整个区域的机动车车道；第二，需要平衡各空间，让整个空间变得和谐；还有需要特别注意的，需要运用当地的材料和植物。

一条循环小径和一条蜿蜒的人行道穿过这个立体空间。在一端有一个文化中心，在中心处有一个运动区域，而在另外一端有一个安静的休息区域，另外还有与周围环境相融的艺术装饰品足够吸引人们的眼球，让他们驻足小憩。文化区域包括四个现代的花园，设计灵感来源于圣劳伦斯河上的四种天气。运动区域包括两个足球场地，空间开阔而且还有两个多功能跑道。休息区域有一个观望塔，一个连接着河流的甲板，还有一个多功能展馆。

景观设计上材料丰富。例如：大卵石、木制配件、柯尔顿耐腐蚀钢阶梯和本土的植物等等。整体的设计风格是现代，简单和优雅。利用熟悉的原木，当地的石头还有河岸原有的树木在现代的韵味上增加了一些亲和力。

Square & Park

广场和公园

Square & Park

Located along the St. Lawrence river, between the Sillery coast and the Ross coast (towards Quebec's bridge), the project delicately weaves a sequence of diverse experiences and atmospheres, navigating from the boundless visual expanse of the river and the scale of the territory, to the tactile sensory experiences of the human scale. The first concern in this project was to inhibit the motorway system which crossed the ground over its length; the second was to try to make a comfortable and immense space by looking after the concepts of scales and balance, but also give a special attention to use local materials and vegetation on the site.

An immense cycle track and a sinuous pedestrian path cross this longitudinal site, bringing the user from a cultural pole at one end to a sport pole at the centre and an interpretation pole at the other end. The journey is punctuated by pieces of contemporary art that interact with the landscape and the urban design. The cultural section of the project gathers four contemporary gardens inspired by the mood of the river——the Quai des brumes, the Quai ds flots, the Quai des hommes and the Quai des vents. The sport section at the center is composed by two ground soccer, big open spaces and a multipurpose track. Composed of an observation tower, a deck connected to the river and a multifunctional pavilion, this section marks an important entry to the site and revives the imaginary of a river pier and its intrinsic structures, turning it into a local visual anchor and light beacon——a significant newcomer to the constellation of industrial relics dominating local landscape silhouette and vistas.

广场和公园

Square & Park

Landscapes are materialised as much with stone boulders, timber assemblies and corten steel thresholds as with native plants and trees, and as with vapour haze, thick shade, mellow light glows and water reflections. The language is contemporary, simple and elegant. The use of familiar material such as rough wood and local stones and the consolidation of the existing shoreline vegetation give an authentic flavour to these contemporary creations.

广场和公园

河滨公园凉亭
Riverside Park Pavilion

项目档案

设计：Touloukian Inc.
结构设计：Richmond So
景观设计：Bob Uhlig
项目地点：美国，马萨诸塞州

Project Facts

Design：Touloukian Inc.
Structural Engineer：Richmond So
Landscape Architecture：Bob Uhlig
Location：Massachusetts, USA

这个项目位于查尔斯河的河滨公园，为来往的人们提供荫凉的休息地及社区聚会。设计的挑战性在于重新设计哈佛大学井口建筑物、排气井以及新建的地下停车场的外部空间。在排气井的外部，设计了一堵木墙、长椅和一个水平的华盖。在设计上，尽量减少对周围环境的影响，创造一个方便社区聚会的公共空间。

A new construction pergola structure for shaded seating and community gathering was built for the new Riverside Park overlooking the Charles River. The design challenge was to conceptualize multiple approaches for re-designing the exterior envelope of Harvard University's elevator headhouse and exhaust shaft that protruded through the site from a recently constructed underground parking garage (parking garage and headhouse structure by others). The shaft's exterior envelope was wrapped by a textured wood wall and bench element and a horizontal pergola canopy was implemented to create a connection to Memorial Drive arid the Charles River, and to minimize the impact of the aluminum louver exhaust system while creating an outdoor community gathering space for the public.

Square & Park

广场和公园

城市甲板
The City Deck

项目档案

设计：StossLU
项目地点：威斯康星州，绿湾
面积：10 117.15 平方米

Project Facts

Design：StossLU
Location：Green Bay, Wisconsin
Site Area：10,117.15 m²

这个项目是一个约 10,000 平方米的长条形河边土地，开发目的主要是为了提高福克斯河的人流量，培养公民意识，提升社会及地方自豪感。项目的第一阶段，意在直接吸引社区成员和市政府官员，以重新激活项目地块。该项目的主要目标是要在新的公共公园和被占领的周围城市街区之间注入一个循序渐进的附属地。几个结构相互关联的设计元素结合在一起，以构造出同时促进朝向和沿着水体两种运动的一个布局。项目中木栈道的曲折和褶皱制造出城市家具的多样性，这使得在一个城市和自然环境中的不同空间有了区别。

条状草坪和硬景观在城市美化项目中交织营造出适合多样活动的空间。灵活的铺地区域特别选在街道入口的地方，刚好适合大型集会。由这些元素组成的不同品质的空间带来极大的灵活性，由使用者自行决定场地的使用规则。

看似简单的形式和材料的逻辑孕育了城市公园爱好者和河流之间的复杂关系。该项目的下一阶段将进一步推动这一想法。护坡台阶、一个舞台、一个城市的长凳和划船码头都将被纳入水体规划中。

Square & Park

The City Deck by StossLU is a long strip of riverfront land developed primarily to increase access to the Fox River and to foster civic and social presence on its edge. The first stage of this multiphase project sought to directly engage community members and city officials in the re-activation of the site. The main goal of the project was to infuse a programmatic dependency between the new public park and the under occupied surrounding city blocks. Several design inter-related design elements are married in order to produce an overall composition that simultaneously promotes movement towards the water and along it. A wood boardwalk twists and folds along the project, producing a diverse set of urban furniture that are placed in zones that promote differing relationships with the urban and natural surroundings.

Slivers of lawn and hardscape intertwine along City Deck creating spaces suitable for many programs. Flexible paved areas specifically sited near street entrances serve as zones for larger spontaneous gatherings or planned events at a civic scale. The diverse quality of spaces created by these elements bring a degree and a flexibility to the project, leaving its ultimate use up to its users.

The seemingly simplistic formal and material logic laid out in the project breeds a spectrum of complex relationships between the park-goer the city and the river. The next phase of this project will push this idea further. Amphitheater steps, a stage, an urban bench and docks for boaters will be incorporated over the water.

广场和公园

北查尔斯顿海军纪念馆
The Greater North Charleston Naval Base Memorial

项目档案

设计：BNIM & Burt Hill 建筑事务所
项目地点：美国，南卡罗来纳州，查尔斯顿

Project Facts

Design: BNIM & Burt Hill
Location: Charleston, South Carolina, USA

这个纪念馆沿着一个视觉性的时间线进行排布设计，它融合建筑、图表和景观，表达着三代海军战船的发展史。设计将建筑与图形穿插其间，为游客提供丰富的体验，使人们能置身于海军的情境中，展示出塑造城市的海军文化。

The memorial is organized along a visual timeline using architecture, graphics and landscape to communicate the evolution of three naval vessels built over the course of the base's operational years: the Landing Craft, the Submarine and the Destroyer. Exemplifying integration between architecture and graphics, this memorial provides an experience for visitors, drawing them into a naval scene and illustrating the naval culture that shaped the city, its people and those who were stationed at the former naval base.

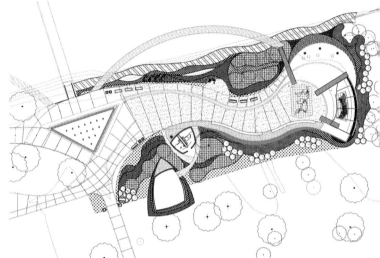

Square & Park

广场和公园

东湖岸公园
The Park at Lakeshore East

项目档案

设计：詹姆斯伯内特办公室
项目地点：美国，伊利诺斯州，芝加哥市
面积：21 043 平方米

Project Facts

Design：The Office of James Burnett
Location：Chicago, Illinois, USA
Site Area：21,043 m²

东湖岸公园是芝加哥内环线东湖岸开发的中心地带，俯瞰着芝加哥河和密歇根湖汇合处。湖滨地区投入40亿美元重建包括4,950户住宅、1,500间客房的宾馆、204 380平方米的商业空间、71 533平方米的零售空间和一所小学。

景观设计师参与了项目早期的总体规划，同时帮助项目组进行开放空间的规划，引导后期的规划和公园开发。融入芝加哥三级交通系统的结果是一项艰巨的改变，需要从南向北偏移25分。两个白色铺装的大型散步区作为从东到西的主要环路，每个散步区设有5个喷泉，使得街区边缘更有活力。儿童乐园位于散步区和北格兰特轴产生的交汇区域，设有一个环形广场，互动流水喷泉和安全的游乐场所。

The park is the central amenity of the Lakeshore East development in Chicago's Inner Loop. Overlooking the confluence of the Chicago River and Lake Michigan, Lakeshore East is a $4 billion redevelopment that will include 4,950 residential units, 1,500 hotel rooms, 204,380 square meters of gross commercial space, 71,533 square meters of retail space and an elementary school at completion. The landscape architect was engaged early in the project by the master plan architects and helped produce the open space guidelines that would later guide the design and development of the park. Integration into Chicago's 3-tiered transit system results in a daunting grade change of approximately 25' from the south side of the site to the north.

Square & Park

广场和公园

菲尼克斯市政遮阴篷
Phoenix Civic Space Shade Canopies

项目档案 | Project Facts

设计：Architekton
项目地点：美国，亚利桑那州，菲尼克斯

Design：Architekton
Location：Phoenix, AZ, USA

菲尼克斯市政公共空间与城市社区以及亚利桑那州立大学共同编织在一起，构建了一个重要的城市中心，与城市新的轻轨系统相邻。波浪起伏的四个绿色遮阴篷构建了一系列的室外空间，反映了公园的波纹地形。这种有节奏的稀松结构是由标准的电子导管和单杆挂钩构成的。切分式的样式和多边的杂色，让这个表面充满了可变性与丰富性。

The Phoenix Civic Space weaves the downtown community together with ASU's Phoenix Campus to create a vital urban center adjacent to the city's new light rail system. Floating above the park, four canopies create a series of outdoor rooms defined by undulating green planes that reflect the park's corrugated topography. This rhythmic scrim is comprised of standard electrical conduit and unistrut hangers: a syncopated pattern and variegated color scheme transform these off-the-shelf components into a rich textural surface that appears to constantly change density as it flows overhead.

Square & Park

广场和公园

Lillejord 休息区及行人天桥
Lillejord Rest Area & Footbridge

项目档案

设计：Pushak
项目地点：挪威

Project Facts

Design: Pushak
Location: Lillefjord, Finnmark, Norway

这个项目是路人的一个停靠点，主要为路人提供长椅、洗手间、垃圾桶和避风的小屋。这里也是通向峡谷中一个瀑布的起点。小径在停车的这一边，一条桥跨过河流，延伸至另外一边。设计师设计的目的就是建造一座新桥，不仅能够起到标志的作用，而且能够满足所有的需求。新的桥将跨过河中央绿色的植被，延伸至一个更为古老的小径上。这样一来，不仅没有打破周围环境的和谐，而且为周围景色增色不少。这座桥并不是直线型的，而是在一端有一个转弯并倾斜下来。这样不仅为路人提供了长椅，还能够腾出建造避风屋的空间。

The program for the road stop is benches, toilet, waste bin and a wind shed. It is also the start of a trail to a waterfall in the end of the vally. The path was originally on the parking side, with a run down bridge crossing the river further up. The new bridge was the designer's proposal; it works as a sign towards the trail, at the same time take care of all the demanded functions. It is leading on to an older trail, crossing the soft, green carpet of vegetation in the midst of the river deltae. By placing all the program in the bridge, the road stop installation is now a distinct object placed in the landscape, which is appropriate for the rough and grand nature of the site. Rather than small furniture placed around or in the ground, the bridge is making a turn, leading to the trail. This also makes room for benches on top of it.

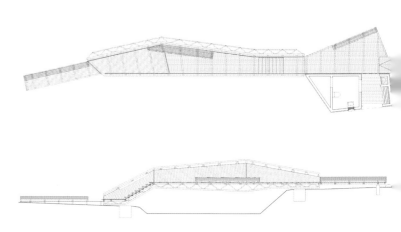

Square & Park

感官花园
Sensational Garden

项目档案

设计：Nabito 建筑事务所
项目地点：意大利
完成时间：2011

Project Facts

Design：Nabito Architects
Location：Frosinone, Italy
Year：2011

这是一个人们可以找到幸福和爱的公园。在这里，人们会增加对彼此的了解，让自己更加融入社区的生活。这个项目的目的在于吸引人们走上一条风景时时变化的小径，让人们自己去感受这种变化，即使是面对着同样的景观。

人类的五官是这个空间的主题，空间所选取的材料和植被都和五官是相关的。随着不断地深入这个空间，人们会慢慢获得各种感官体验。在这个空间内，高度和倾斜度的多样性使得花园更加有趣。设计师们利用这些感官含蓄地将他们自己，人们和周围的环境联系在一起。沿着小径，每个小空间逐一展开，五个形状不同的混凝土突起物代表着人的五官，有一条小径蜿蜒在其中。在这里，你可以闻到清新的味道、可以听到各种悦耳的声音、能看到漂亮的玫瑰花园、触摸到各种材质、品尝到美味的水果。

It is a garden in which users and citizens can find the joy of life, love and get to know each other again and make themselves comfortable with the entire neighborhood. The goal of the project is to invite users to a path in which the scene is always changing. The user will have the sensation of discovering an always-different space but with the same kind of characteristic. The five human senses are the main theme of the space, the materials and vegetation will relate to them. The user will not have an entire look over the park, but he will do through a series of the different senses. The variation of height, inclination, and dimensional games is part of the ludic peculiarity of the park. Nabito Architects and Partners use the senses as a big metaphor. They use senses to relate ourselves with surroundings and other people. Each area is a metaphor of one of the five human senses and you are constantly called to put yourself in relation with the space. The path is a discovery, and it is designed to leave the spaces to be revealed to a visitor little by little, in order to induce and encourage the visitor of continuing the experience. Five Big Devices contain the essence and the poetry of the metaphor. A path is the link between them. The smell is attracted by the support of the essences: hearing, from the game sound amplification, view, from the beautiful rose garden, touch, from feeling the materials of the central cone, and finally, taste stimulated by the fruit trees in the largest support.

广场和公园

广场和公园

NPS 屋顶花园
NPS Podium Roof Garden

项目档案	Project Facts
设计：PLANT Architects 建筑事务所	Design：PLANT Architects
项目地点：加拿大，多伦多	Location：Toronto, Canada

这个平台上的屋顶花园将上层定义为一个公共花园，融合高抬的走道系统，同时设计也尊重综合体的历史背景，将一个真正的二十一世纪交流空间的形式呈现在人们面前。尽管这是加拿大最开放也是最方便公共游览的屋顶空间，包含着各种技术环境优势，但在设计时它只是定义为一个巨大的花园，包含复杂的色泽和错落有致的材质。景观的安排突出了建筑的形式和材质，为人们提供了休闲娱乐聚会的好去处，同时也在周围高层中形成强烈的视觉效果。

The Podium Roof Garden re-conceives this upper level as a public park integrated with the elevated walkway system, respecting the complex's heritage status and reopening it to the public as a truly engaging 21st century space. Although it is the largest publicly accessible green roof in Canada providing a plethora of technical environmental benefits, the Podium Roof Garden is designed as a vast garden park which bears close scrutiny of its complex color and textural mosaic. It is structured to closely reveal the compelling form and texture of the buildings, and provides plenty of spaces for strolling, lingering, intimacy and gathering. Simultaneously, it provides a compelling graphic legibility from the surrounding high-rise buildings.

Square & Park

广场和公园

Perruquet 公园
Pinar del Perruquet Park

项目档案

设计：ARTEKS 建筑设计事务所
地点：西班牙，塔拉戈纳省
面积：28 000 平方米

Project Facts

Design：ARTEKS Arquitectura
Location：La Pineda (Vila-Seca), Spain
Site Area：28,000 m²

在地中海周边的沿海地带，由于海上浮油的污染已经严重影响到了树木与其他植物的生存与生长，因此，单纯的创建像自然森林一样的场所是不可能的，而这个 Perruquet 公园则是 ARTEKS 建筑设计事务所参照树木的形状设计的纯建筑形式的休闲空间。其中最重要的一个特点就是柱子的运用，海边树木的一个最突出的特点就是略向强风一方倾斜，而这种强风也决定了重新栽种树木是不可行的，因此，就用这种柱子的垂直性与海平面形成一种交叉对照。公园中的棚子采用玻璃纤维制作的六角形网格结构，玻璃纤维可抵制海盐的侵蚀，它还带有波浪形，可以随风轻轻移动。

In Mediterranean peri-urban littoral, the pollution of maritime aerosol affects the survival and the growth of pines and other plants. In this way, creating an extension forest like the existent one is not viable! One of the most characteristic signs of the littoral pine is its light inclination in direction of the dominant wind. This wind is the cause of the impossibility to replant pines. This vertically contrasts with the horizon line of the sea. The canopies are made by a hexagonal grid of fibreglass (material resistant to the marine salt) that moves lightly with hard wind. Mounted like a "Mecano" very fast, the time is spend in design.

Square & Park

广场和公园

Square & Park

广场和公园

伊利街广场
Erie Street Plaza

项目档案　　　　　　Project Facts

设计：StossLU　　　　Design：StossLU
项目地点：美国　　　　Location：Milwaukee, Wisconsin, USA
面积：13 000 平方米　　Site Area：13,000 m²

这是位于 Federal 河道与密歇根湖之间的一个新的公共空间，目的是为了连接 Milwaukee 商业区和新复兴的城市郊区。它是三英里长的绿色走廊的一部分，旨在推动沿城的活动和对滨水空间的公共利用。从项目开始，它的形式就在最大实用性、环保性和推行生物多样性之间不断发展变化，同时设计通过三种元素的使用对场地的工业过去也有一定的反映。将公园看成一个灵活的场地是设计的重点，设计师从这一点出发创造出一种能与最终功能相匹配的设计。

通过对三种元素的操作或是对生态性的复合，设计实现了美感和建造的多样性，这三种元素分别是辐射状的树林、灵活的场地及从场地边缘蔓延至水边融合了钢铁元素的沼泽。每个独特的区域都有不同的项目元素，景色和生态特征。

Positioned on the edge of the Federal Channel and Lake Michigan, Erie Street Plaza by StossLU is a new public space that serves to connect downtown Milwaukee to a newly revitalized outer zone of the city. It is part of a three-mile-long green corridor that promotes activity along the city and reclaims waterfront for public use. From the project's inception, its form progressed under the principle of maximizing programmatic, environmental, and ecologic variety within the space while recalling the site's industrial past and surroundings through use of three primary elements.

Viewing the park as a "flexible field" was key for StossLU in creating a design that could behave in a way that would parallel its ultimate use. That being said, aesthetic and built variety is attained within the park by the formal manipulation of three elements or liybrid ecologies. The radiant grove, the flexible field, and the steel marsh step down from the edge of the site towards the water. Each distinct zone houses varying program, views, and ecologic characteristics.

Square & Park

广场和公园

美国迈阿密南岬公园
South Pointe Park

项目档案

设计：Hargreaves Associates, Inc.
地点：迈阿密，美国
面积：89 000 平方米

Project Facts

Landscape Architecture：Hargreaves Associates, Inc.
Location：Miami Beach, Florida, USA
Site Area: 89,000 m²

这是一个充满生气的生态公园。原来的公园只有77 000平方米，经过重新设计和改造，如今的公园有89 000平方米，有两条环形的小道穿插其中。这两条小道并不是紧贴在地面上的，而是有着一定的高度。这样，就增加了整个空间的层次感和立体感，并将地面分成若干部分，与地面有明显的界限。远远看上去仅仅是两条存在的线型结构，但当人们走在这两条小道上却获得了无与伦比的亲身体验感受，一块珊瑚的化石将临近海中自然进化的过程呈现在陆地上。独特的地形可以增加人们对活动的选择性。公园靠近海岸的一侧种植的是本土的沙丘植物，而内侧种植的则是棕榈树。另外，在公园内还有一个小型的花园，里面种植着本土的地被植物、棕榈树和落叶性树木等。

公园有一个凉亭，凉亭内有提供咖啡和休息的设施。带有坐台的竞技场成为一个小型的舞台，方便人们日常的聚会和活动。而且，在这里人们可以欣赏到远处的海景。公园的另外一个亮点就是三个公共的草坪，人们可以自由进行各种活动。在草坪上种植的是抗盐侵蚀的矮草，这些草可以保存水分。在建造过程中移出的树木被保留着，然后根据设计，最终栽植在草坪上。在公园刚刚对外开放的时候，这些起到了巨大的作用。专门定做的夜灯不仅为人们创造了美丽的夜景，为巡航的船只提供了照明，也为海龟们每年一次的返回沙滩增色不少。在海龟们为期六个月的筑巢过程中，灯光将会被调成琥珀色，不会影响到海龟们的繁殖和返回海里。

South Pointe Park creates an animated and ecologically sensitive community park in lively and flamboyant South Beach. The park is a redesign of an existing 77,000m² park along Government Cut. The now 89,000m² park positions two corresponding circulation paths. The height of the path gives it the heft of a sculptural object and sets it apart from the surface of the park. Constructed from Dominican Keystone, a stone of fossilized coral suggests the natural processes of the adjacent sea. From the park, the path is perceived both as object and line, while the experience on the path heightens the sense of spectatorship.

The landform encourages spirited movement along it, and the heightened experience of an ever-changing visual field of movement is enlivened and enthralling. The twisting of the landform is echoed as a motif throughout the park in smaller garden areas. Sinuous bands of native dune plantings on the ocean side of the serpentine landform are contrasted with abstracted dune landforms and palm trees on the inside slope of the serpentine. A smaller garden area of coastal hammock plantings of native ground covers, palms and deciduous trees echo the twisting path of the landform. At the interior of the park, a pavilion with cafe and facilities creates a point of rest. A seatwall amphitheater creates an informal staging area and seating overlooking the water playground. The park also features three areas of open lawn which support free and flexible program. These areas of lawn are planted with salt tolerant turfgrasses to withstand active use and are engineered to retain water after tropical storm events. Mature trees from the site were stored during construction and then replanted on the lawn, achieving an immediate visual impact at the opening of the park.

Custom site lighting was designed to provide a signature night-time experience and announce the point of passage to cruise ships, and they signify the annual phenomena of sea turtles retiming to the beach, and to nest.

Square & Park

广场和公园

Freres-Charon 广场
Square des Freres-Charon

项目档案

设计：Affleck + de la Riva 建筑事务所
项目地点：加拿大，蒙特利尔

Project Facts

Design：Affleck + de la Riva Architects
Location：Montreal, Canada

项目采用一种简单而精致的方式，用最低限度的建筑语言构建了一组以圆形和圆柱形为主的野趣花园，花园边缘处有一个风车造型的望景楼。为了与花园相互映衬，照明系统采用了彩色变换灯光，以适应四季的变化。这个项目为当地居民提供了一处宽敞舒适和自豪的公共休闲空间。

设计师们从现代生活方式出发，站在使用者的角度，将广场融入到周围的环境之中。这个广场安全、舒服、可达性很强，即使是坐轮椅的人也可以自由进出。栽植的植被也主要是当地植被，节省了大量复杂的灌溉系统。设计师们通过利用当地的植被和回收的亭子减少了项目的成本，为居民和当地的旅游产业带来了利润。尽管这个项目的预算和空间都是有限的，但这个广场却是城市主要街道里最具有历史韵味的，也是蒙特利尔旅游业发展的支柱。

The project uses a simple, refined, and minimalist architectural language to create a dialogue between circular and cylindrical forms including a garden of wild grasses, the vestiges of the windmill and a park pavilion in the form of a belvedere-folly. Complementing these gestures, the lighting scheme proposes a chromatic garden that alludes to the changing seasons.

By focussing on the experience of the contemporary city and urban lifestyles, the design team explored concepts from a user's point of view and initiated a connection with the immediate surroundings. The square's street level public domain was carefully designed to insure it comfortable, safe, and wheel-chair accessible. Sustainable initiatives include the planting of local species of wild grasses which take a significant load off the municipal irrigation system.

Affordable amenities such as an interpretive program, extensive vegetation and a recycled park pavilion allowed the designers to control costs while providing significant benefits to residents and the local tourist industry.

While modest in scale and budget, the square is an essential component of McGill Street's larger network of historic spaces and a key element in Montreal's cultural tourism branding strategy.

Square & Park

广场和公园

Square & Park

广场和公园

Ricard Vines 广场
Ricard Vines Square

项目档案

设计：Benedetta Tagliabue 建筑事务所——Josep Ustrell
项目地点：西班牙，莱里达
面积：9 200 平方米

Project Facts

Design：Benedetta Tagliabue——Josep Ustrell
Location: Lleida, Spain
Site Area：9,200 m²

这个项目的所在地是莱里达最美丽的地方，项目围绕着 Seu Vella 大教堂，主导着整个城市。这个广场的设计焦点是创造一大片绿色空间，以突出为音乐家 Ricard Vines 打造的雕塑。场地内充满了大大小小的空间，形成不同的迷宫，营造出神秘的远古氛围。设计师们意图建造一个开放的"跳舞毯"式的地面，并通过构成迷宫的小径来指导人们跳舞。

这些小径构成一个大圆环，圆环是由植被和砖块铺设而成的。新的广场被分割成很多小的绿色区域，依稀可见树木和灌木丛。这些绿色区域不仅为孩子们提供了游玩区域，也为行人提供了带座椅设施的休息区域。

Square & Park

The large green open spaces that surround the Seu Vella Cathedral and dominate the whole city are the most beautiful public areas in Lleida, and are what we chose as our reference point. The designs for the new Ricard Vines Square must possess some of this beauty. The focus of the proposal is to build a large green open space for a sculpture dedicated to the musician Ricard Vines, and a space full of little squares cross section and green areas at a point where the city throngs with traffic and pedestrians. The maze or labyrinth provides an ancient model. Architects propose an open space featuring a dance floor with a labyrinthine path guiding the steps of those dancing the spring dance around the central feature that generates and guides the movement of the dance, filling the surround space with life.

The project is made up of a circle and new square for the city. The circle ends in a labyrinthine pattern, made up of bands of planting and brick paving. The square is split up by green areas; a maze of trees and low shrubs are visible from the roadway. This greenery helps organise the neighbourhoods leisure activities with bars, children's play areas, benches and pedestrian pathways.

广场和公园

爱沙尼亚路博物馆的露天展厅
Open Air Exhibition Grounds

项目档案	Project Facts
设计：Salto AB	Design: Salto AB
项目地点：爱沙尼亚	Location: Varbuse, Estonia

这个露天博物馆是沿着一条大路形成的。经过时，路线上会出现很多不同的景观，沿路包括很多八字形的排布，其中点缀不同的展品，如同连环画一般。博物馆空间还包括一些从地面掏空的圆圈，它们与周围的自然景观形成了明显反差。

The concept of the open air exhibition grounds is based on a road. While passing by, your route will be surrounded by different landscapes. The solution forms a long 8-shaped path, where functions with different character are placed in succession like a comic strip. All space necessary for the museum is scooped into the landscape leaving rest of the environment as natural as possilole; natural and artificial landscape is clearly separated. A hollow ranging from 10 cm to 4 m forms largescale open-air exhibition grounds.

Square & Park

海景廊架
On the Way to the Sea

项目档案

设计：Derman Verbakel 建筑事务所
项目地点：以色列，Bat-Yam
完成时间：2011

Project Facts

Design：Derman Verbakel Architecture
Location：Bat-Yam, Israel
Year：2011

这个项目的设计宗旨是鼓励游人在海岸活动，并且体验从城市到海边这个路程的乐趣。一系列的方框门和带有轮子可动的椅子，桌子等在设计中诞生。游客可以自由地移动这些桌椅装置，根据需要随意组合；在某种程度上来说这些实现了在公共区域打造个人空间的愿望。灵活的景观设施和桌椅等功能设施，激发了游客海边游玩的乐趣，提升了该区域的休闲性。

在项目和海滩的交界处，设计了一个开放式的露台。在这里，人们可以乘凉，并欣赏海景。项目不仅满足了人们的日常生活需求，而且还能满足特别节日时候的需求。

Square & Park

The way to or from the sea passes through the site but the movement from point A to point B is not the purpose. A series of frames carefully positioned between city edge to sea shore host public activities, creating a new use for this space. The installation invites inhabitants and passers-by to intervene and create opportunities for events and unexpected interactions by manipulating different elements integrated within the frames.

At the interface between the project and the beach, an open terrace offers views to the sea, providing shade and reclined seating facing the horizon. Together, the elements create a micro-climate where people can meet, play, eat, talk or just hang out, thereby producing a platform for a wide range of possible interactions, from daily uses to special events.

广场和公园

9女孩纪念馆
M9 Memorial

项目档案

设计：Gonzalo Mardones Viviani
项目地点：智利，圣地亚哥
面积：160平方米
完成时间：2011

Project Facts

Design：Gonzalo Mardones Viviani
Location：Santiago, Chile
Site Area：160 m²
Year：2011

这个项目位于地下，中部向着天空开了一个直径约3米的锥形开口。9种光源，9个区域，象征着9个女孩。灯光非常神圣，增添了混凝土建筑的吸引力。这是一个将天空也吸纳了的空间，它以人性化的姿态，友好地欢迎人们的到来。人们可以在这里聚会，休息和沉思。

一个长16米，坡度为12%的坡道将人们引向这里。中心处有一棵木兰，这种树木非常特别，是唯一的一花九叶植物。地面采用石块来铺设，而入口处倾斜的表面则用草来铺设。这种绿色坡道入口使用了嵌在混凝土里的钢化玻璃作为保护，一侧嵌入了27盏灯。

Square & Park

The work is constructed from the underground and opening up to the sky through a concrete cone of nine feet in diameter. 9 lanterns and 9 modules symbolize the 9 girls who left earlier. The presence of divine light allows gravitate concrete forms. The folds of concrete are shown as a human gesture with which the architecture welcomes a gathering place, a place of refuge, a meeting place within the park.

It is accessed by a ramp 16 meters long with a 12% slope to reach the memorial oratory. A magnolia is planted in the geometric center. This tree is the only one of nature that has 9 leaves into each one of its flowers. The floor is paving stones. A protection guardrail system uses a tempered glass embedded in concrete elements. A band of concrete on the paved surface to create a bench faced with the presence of the virgin welcomes 9 candles and 27 smaller candles which are inserted into the folds of concrete.

广场和公园

Square & Park

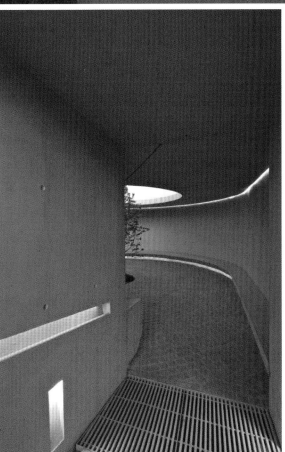

Holding Pattern 公共装置
Holding Pattern, PS1 2011 Installation

项目档案

设计：Interboro Partners
地点：美国，纽约
完成时间：2011

Project Facts

Design：Interboro Partners
Location：New York, USA
Year：2011

这是由 Interboro Partners 在今年 PS1 年轻建筑师项目中设计的获奖项目，它是一个简单的大型公共装置。这是一个以社区为基础的设计，设计中有两个要求：材料必须是可回收的；乒乓球桌、凳子和投光灯等共 79 件公共设施捐献给长岛社区内的相关机构。

Holding Pattern is a community-based design that incorporates both the program requirements and the communities needs. Stretching the funds, Interboro Partners will be able to serve two purposes: the materials will be recycled; donating ping-pong tables, benches, and flood lights (a total of 79 items) to over 50 organizations in the Long Island neighborhood. Recycled plywood, sandboxes, both ping-pong and football tables, even a lifeguard stand and wading pools current fill MoMA's well known PS1 courtyard space.

广场和公园

 Dzintari 森林公园
Dzintari Forest Park

项目档案	Project Facts
设计：Substance	Design：Substance
项目地点：尤尔马拉，拉脱维亚	Location：Jurmala, Latvia
面积：131 108 平方米	Site Area：131,108 m²

Dzintari 森林公园是独一无二的，因为它位于尤尔马拉市中心。项目占自然基地 131 108 平方米。由于这个地区公共和住宅密集发展，将这个森林公园纳入到城市的基础设施整体系统中，使之成为人们常去的地方，是项目的重点。一株 200 年树龄的松柏和越桔灌木丛是森林公园最大的财富，因此，新的基础设施构建了一种独特的机械装置，保护古树的同时使游客最近距离地和自然接触。

模块系统的形式原则最适合该公园建筑。模块像树枝和根一样分叉，保护周围的自然的基础元素，并按照类似的原则发展出道路结构。因此，建筑呈现角度不同的平面和立面，大大减少了整体笨重的体量。复合板与表面抛光铝用于外立面表皮，镜面般垂直的表皮反射着周围的自然景观，减弱了建筑对环境的影响。

Square & Park

Dzintari forest park is unique due to its location. Its 131,108m² territory of nature base is located in the very centre of Jurmala city. Due to intensive development of public and residential objects around this nature base territory, the idea to include it into the overall system of the city's infrastructure objects and to adapt this territory to regular public visits became topical. A 200-year-old pine-tree growth and the protected biotops of bilberry bush are the greatest treasures of Dzintari forest park. The new infrastructure is created as a singular mechanism that controls relationship of park and visitors.

Modules system is selected as the most suitable principle for form-creation of the park's building objects. Modules ramify like tree branches or roots, go around protected nature base dements and develop into foot-path structure that was created based on similar principles. As a result, facades of building objects are broken into separate planes in different angles that significantly reduce the overall bulkiness of volumes. Composite panels with polished aluminium surface are used for facade finishing. Vertical division of specular facades reflect the natural appearance of surronding environment and dispel the park's newly created building objects.

广场和公园

夏洛特花园
Charlotte Garden

项目档案

Project Facts

景观设计：SLA
项目地点：丹麦，哥本哈根
面积：13 000 平方米

Landscape Architecture：SLA
Location：Copenhagen, Denmark
Site Area：13,000 m²

这个花园位于哥本哈根一个住宅区的中央空地上。在这里，植物的外形、生长和色彩变化成为花园的主要特色。我们主要采用了各种粗放管理的草本植物，如那些当地草种、兰羊茅、巴尔干蓝草和紫色酸沼草。花园的形态主要取决于各种草本植物造景及植物的特点。可以说，决定这个住宅小区空间面貌的不是建筑物，而是各种植物和植物的生长变化。之所以想到使用野草，首先是因为这类植物可以创造出各种不同的空间，草本植物本身并不稀奇，只有在整体运用上才会出彩，所产生的效果也不仅仅只是装饰性的。

The planting consists mainly of different grasses such as meadow grass, Festuca glauca, Seslevia and Molina caerulea. Unusual for Scandinavian latitudes there is now colour all year round. Colours that change from blue and green in the summer to golden tones in the winter——a space of nuances. The different and changing spaces are held together by paths crossing through the garden, whilst the delineation of the spaces is achieved by means of change of material. A textural and sensory space pays a particular attention to nuances and movement.

Square & Park

广场和公园

Arboretum Klinikum 花园
Arboretum Klinikum

项目档案

设计：Idealice
位置：奥地利，卡林西亚
面积：60 000 平方米

Project Facts

Landscape Architecture：Idealice
Location：Carinthia, Austria
Site Area：60,000 m²

奥地利 Klagenfurt 医院被分为一个开阔的公园场地和多个小院落。西侧、南侧和北侧的公园在河道和其他十九个院落之间形成过渡空间。根据当地自然环境，一些河边的野生植被被种植在公园和院落里，而南侧的院落则是装饰性的植物品种。开阔的空间内系统地分布着不同层次的植物品种和不同类型的灌木品种。除了树和灌木之外，还有苜蓿的融入，避免了过敏的发生。每个院落都有独特的色调，这为游客、病患还有工作人员提供了更好的方向指引。

院落最大的特点就是种植了大量的树木品种，例如：橡树、枫树、樱桃树、柳树、野果树。小山的形状设计来源于不同树皮，这种设计理念贯穿于整个花园设计中。

Square & Park

The grounds of the regional hospital in Klagenfurt are divided into extensive parkland and courtyards. The western, southern and northern parks form the transition between the river area and all of the 19 inner courtyards. Based on the natural habitat, wild varieties of tree species (a different leading tree species in each yard) typical for the river landscape, are planted in the northern park and courts, while in the southern courtyards decorative plants are used. The open spaces of the LKH Klagenfurt NEU are called "Arboretum Klinikum" due to the thematically structured collection of tree species and bush species of different origins. The choice of trees and shrubs is made jointly with medics in order to avoid allergies. Each courtyard prevails a different flowering colour that offers visitors, patients as well as the staff better orientation in the hospital complex.

The courtyards also feature different leading tree species such as oak, maple, cherry, willow, lime and wild fruit. The forms of the hills are based on the shapes of different tree barks, that were used in the concept of "Aboretum Klinikum".

广场和公园

Gleisdreieck 公园
Park am Gleisdreieck

项目档案　　　　Project Facts

设计：Atelier LOIDL　　Landscape Architecture：Atelier LOIDL
项目地点：德国，柏林　　Location：Berlin, Germany
完成时间：2011　　　　Year：2011

这是由 LOIDL 工作室设计的 Gleisdreieck 公园，公园位于德国克罗兹堡中心，原来是一片铁路场地，几十年来一直处在被人遗弃的状态，这是首次被设计师发现并重新使其焕发生机，重新融入到城市的结构中来。开阔的延展和高原一般的场地带给人不一样的感受，加上之前老铁路基地的氛围，Gleisdreieck 公园无疑可以说是柏林独一无二的。景观设计上注重基本元素的运用，没有任何夸张的修饰。空间简单形成，并通过恰到好处的细部、材料以及植被将本身所具有的潜能充分发挥出来。

The park is in the heart of Kreuzberg, a former triangular junction and a waste railway land. Now, for the first time, this tract of land has been reintegrated into urban structure. The impressive expanse and the other-worldliness of the plateau-like grounds which lie four meters above city level as well as the relicts of the former railway junction make the park unique in Berlin. The park restores landscape architecture to the basic essentials. Without any decoration, urban spaces are formed, unfolding their full potential through fine details, sensual materials und vegetation.

Square & Park

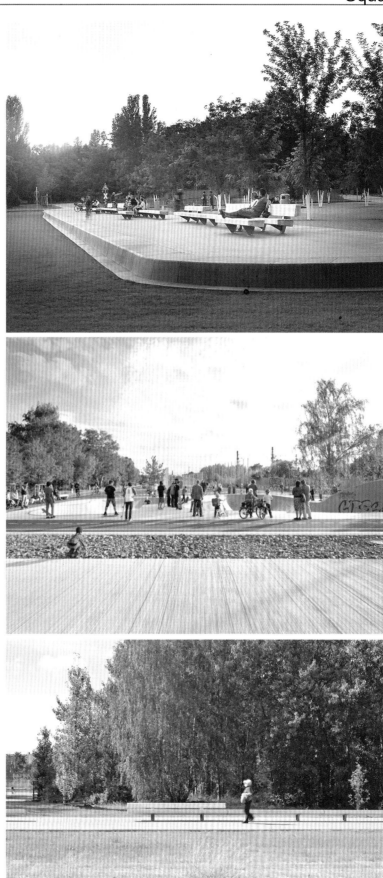

广场和公园

万桥园
Garden of 10,000 Bridges

项目档案	Project Facts
景观设计：West 8	Landscape Architecture：West 8
项目地点：中国，西安	Location：Xi'an, China

"万桥园"试图效仿自然山水带给人视觉、尺度和心理上的冲击。借助地块邻湖且较为中心的地理优势，吸引参观者的将不仅仅是园林本身，还包括世园会的山水美景。设计的概念非常简单有力：桥、小径和竹子。

一条单方向的砾石小径只设一个入口和一个出口，贯穿全园。狭窄的小径象征着生命，如古希腊神话中的名匠Daedalus设计的迷宫，曲折盘旋，带领参观者远离宽敞的大道和人群，深入茂密幽深的竹林，经过每座桥的桥上和桥下。参观者身处园中，不知自己所处为何，也无法看见前路还有多远，"只在此山中，云深不知处"。唯一可以清楚把握的，仅仅是此时此地的自己，或许还可以隐约听见透过竹丛传来的，在其他小径上探索前路的隐隐人声。"空山不见人，但闻人语响"。这两句唐诗，很好地概括了中国人的自然观，也概括了我们想通过万桥园展现给参观者的自然景观。

所有的迷局和不解在五座彩虹般的拱桥上得到了解答。当参观者沿着狭窄陡峭的楼梯艰难地登上桥顶，一目了然的不仅仅是已远在脚下的竹林、周围雕梁画栋精彩绝伦的园林，世园会的山山水水都尽收眼底。每座桥都是参观者的必经之路，它们的位置经过精心选择，如桂林的山峰层层叠叠，也和世园会的总体布局相呼应，使参观者站在每座桥上都能得到不同的视角和景观体验。

"The Garden of 10,000 Bridges" plays with perspective views, limits and the sensation of surprise. The integration of views to the surrounding parts of the exhibitions takes advantage of the perfect position of the plot towards the lake and main features of the park. The concept is very simple and strong. There is only one entrance and one exit of the path which goes through the garden. The narrow path is curling through the garden of bamboo and passing over and under each bridge in the meanwhile. The visitor is not able to see at which point of the garden he is located and how much of its way he has finished. In these moments the visitor is limited to himself, possibly the person in front of him and the sounds of the moving bamboo.

Square & Park

The situation changes dramatically at the 5 bridges. The visitor climbs up the steep bridges to reach the top that sticks out from the dense vegetation. Each bridge allows another spectacular view. The bridges are situated to fit perfectly into the optimal relationship to the viewpoints that are defined by the surrounding.

广场和公园

格拉纳达论坛广场
Forum of Granada

项目档案

设计：Federico Wulff Barreiro, Francisco del Corral
项目地点：西班牙，格拉纳达
面积：11 000 平方米

Project Facts

Design：Federico Wulff Barreiro, Francisco del Corral
Location：Granade, Spain
Site Area：11,000 m²

这个广场位于格拉纳达的城市边缘，周围都是无限的农村风光。这个项目的主要目的就是将传统的农村景观融入到现代建筑当中。山脉、城市，新的文化设施定义了新的景观。
整个地形灵活性很强，不仅可以指引观光者的移动路线，而且给广场带来了新的生命力。地面上所采用的材料是不同的，主要突显自然的特点。广场在空间上分为上下两个层次，第一个层次就是入口；第二个层次由植被斜坡延伸下来，用于各种商业和社会活动。整个地形时而连续，时而断开，并通过明显的线条让观光者亲身体验这种变化。这些线条分隔了不同的树木和灌木丛区域以及水池。
所选择的材料提升了整个项目的质量。地面主要采用木板、白色的混凝土和黑色的石头。线条是由柯尔顿耐腐蚀钢构成的。入口处，简单的几何形状和香气，定义了整个空间。在低层里，多种颜色、质地以及灌木丛延伸了整个空间。

Square & Park

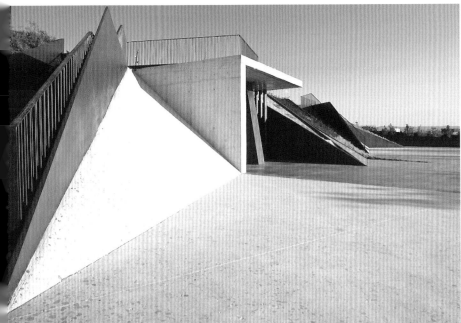

In the limit where the city edge of Granada merges with the agricultural landscape of its surroundings, the new Forum public space is developed. The project pretends to establish a dialogue from a contemporary perspective with the traditional agricultural landscape preserved from the new building construction. The visual relations with the Sierra Nevada mountain, the historical city, and the new cultural equipment devdoped in the area, defines the new landscape.

This new territory is defined by a flexible intervention system, which permits to direct the movement of visitors and improve the perception of the new qualities of the site, as the introduction of interchangeable uses. In ground's level, the work has been conceived as a series of diffracted bands of different materials. Between them, the nature emerges. The garden is developed into two levels. The first one is the access, and the second, at a lower level, sheltered from the highway by elevations of the vegetal bands, allows the development of business and social events. This territory of continuous and broken bands is a garden defined by force lines that follows our walk and invite us to feed it. This is an area of convergence, based on lines that define the areas of different kinds of trees and bushes, and the whisper of water in a pond.

广场和公园

The geometry of the project defines its construction. The chosen materials enhance the qualities of this new territory. The floor is made by warm bands of wood, white concrete with black stones, a contemporary evocation of an historical pavement from Granada, and dark slate. The force lines that define this bands are constructed in corten steel, and rise from the level of the ground to lead, protect, or contain. In the access, the simple geometry and the perfumes define the spaces. On the lower level, color and variety of textures and volumes of bushes extends the spaces.

广场和公园

Las Arenas 广场
Las Arenas Square

项目档案

设计：Javier Perez
项目地点：西班牙
面积：22 000 平方米

Project Facts

Design：Javier Perez (ACXT)
Location：Getxo / Vizcaya / Spain
Site Area：22,000 m²

这个项目的目的是将两个不同社区的广场成功地连接起来。这两个不同的广场被一条铁路分隔开来。虽然地下广场的建造消除了地理上的界限，但是心理上却还是有界限。而将两个社区，两个广场，两个空间连接成功，应该是不管在地下还是在地上都是要很好地相连起来的。

The aim of this proposal is to achieve a successful link between the two squares belonging to two different neighborhoods. They used to be separated by a railway line. The construction of the Bilbao underground eliminated this physical barrier but a psychological division still remains. Reunite two neighborhoods, two spaces, two squares: down on the ground, up in the sky.

Square & Park

广场和公园

 Cufar 广场
Cufar's Square

项目档案

设计：Bruto Matej Kucina, Domen Sega
项目地点：斯洛文尼亚，耶塞尼采
面积：3 000 平方米

Project Facts

Landscape Architecture：Bruto Matej Kucina, Domen Sega
Location：Jesenice, Slovenia
Site Area：3,000 m²

这个广场正好是耶塞尼采城市生活的中心。白色的条带、数个电脑控制的喷水器为广场增添了动态之美。周边还有中学、电影院、歌剧院和图书馆。这个广场拥有了所有的城市广场的功能。

In the very centre of Jesenice urban life, surrounded by a high school, cinema, theater and library, lies Cufar's Square. Simple, yet very dynamic urban intervention, designed by Bruto and Scapelab, Cufar's Square provides all a city square should offer——social and multifunctional event space. White stripes and a pack of computer-controlled water jets add a layer of play and dynamic to the square.

广场和公园

里昂草地公园
Lyon Meadows Park

项目档案　　　　　　　Project Facts

设计：BASE　　　　　　Landscape Architecture：BASE
项目地点：法国，里昂　　Location：Lyon, France
面积：60 000 平方米　　 Site Area：60,000 m²

这座草地公园位于 La Chapelle St-Luc 镇，属于法国特瓦鲁的郊区。这座公园最初设计于 70 年代，是英式公园的模型。但随着时间的流逝，公园逐渐被荒废，因此市政厅决定对这座公园进行翻修，使其变得更加迷人。

BASE 设计工作室通过对公园内的土木工程进行复原、道路修缮、入口重新设计以及多植物区的打造，使该公园变成了年轻人最常光顾的区域。

设计师为公园打造了面积为 700 平方米的"混凝土"街道，同时还有儿童游乐场、迷你高尔夫球场、地球游戏场以及健身小径等等。公园的中心是一片广阔的平地，BASE 将其打造成面积为 170 平方米的日光浴室，使之成为人们休闲放松的好去处。

Located in the town of La Chapelle St-Luc, on the outskirts of Troyes, the park is a city park which was designed in the 70s on the "English garden" model. This park suffered numerous degradations so the town council decided to contract landscape architects to make it more accessible and attractive.

BASE's work consisted in reinstating earthworks, pathways, entrances and most planted plots, proposing several play areas, most of them designed for the young.

A 700 m² concrete "street" skate park has been built along with children playgrounds, a miniature golf course, peatanque playing grounds, a health and fitness trail.

At the centre of the park, a vast plain liable to flooding remained dry and unused. BASE proposed the building of a 170 m² solarium facing south destined for sunbathing and designed more largely as a chill-out and relaxation area. Access is through a network of wooden pontoons and banks, located 1m high, safe from potential flood risks. Integrated lighting is provided in the structure and pontoons.

Square & Park

广场和公园

公共图书馆和阅读公园
Public Library and Reading Park

项目档案

设计：Martin Lejarraga
项目地点：西班牙
面积：18 500 平方米

Project Facts

Landscape Architecture: Martin Lejarraga
Location: Spain
Site Area: 18,500 m²

这个项目充分利用了城市特殊的地形，并赋予其新的面貌，为居民提供了文化与娱乐为一体的公共空间。整个地面呈折叠状，地面上的图书馆和公园经过严格的位置设计，创造出新的空间，便于人们逗留与交流。图书馆在公共空间之中，同时也被这个空间保护着。这个项目还充分利用了特别的地中海环境以及具有丰富文化的社会环境，新建筑与周围环境的兼容处理得相当漂亮，还意外创造出其他更小的空间，例如：公共大厅、会议室、运动场、花房、游玩区等等。

The project comes out of the creation of a new topography that indexes and qualifies this zone of expansion in Torre Pacheco, on a plot of public equipments——a urban, cultural and enjoyment alternative for the citizens. The folded plans of the area characterize the performance, in which both equipments that occupy the plot, library and park, adapt their relative position, creating new spaces protected from reception, communication and stay. The public space contains and protects the building——"two faces of the same coin". In this case, completing the built program with an open air park takes advantage of the special conditions of the environmental (Mediterranean, very soft) and social (of exchange and multiethnic relation) climate of a prosperous and growing city such as this. The possible and desirable open relationship of compatibility, optimization of spaces and services, between the new building and the surrounding ground, generates a wide disposition of different general areas of common use between both of them(library, public hall, meeting rooms, etc.) and urbanized spaces (sports courts, greenhouses, gardens, playing areas, swings, etc.) that widen the real used and enjoyed space for the entire city.

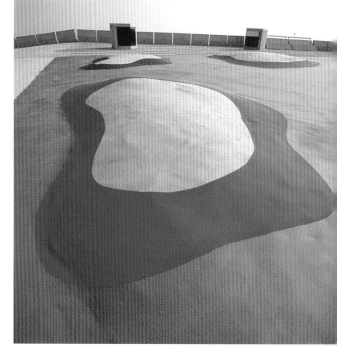

Square & Park

广场和公园

Asmacati 购物中心
Asmaçati Shopping Center

项目档案

设计：Tabanlioglu 建筑事务所
项目地点：土耳其，伊兹密尔
面积：22 760 平方米

Project Facts

Architects：Tabanlioglu Architects
Location：Izmir, Turkey
Site Area：22,760 m²

Asmacati 购物中心和聚会中心位于伊兹密尔的城市中心，这个中心赞美并歌颂了伊兹密尔城市的室外生活方式，同时又融合了当地温和的气候条件。半开放的商店设施自然地在商店之间形成娱乐区。而由自然材料做成的凉亭为人们提供了休闲和遮荫的地方，凉亭设计模仿的是当地景观中的葡萄藤叶的形状。

Asmacati Shopping and Meeting Point is located in the city of Izmir. The center appreciates and joins the lifestyle of Izmir where people prefer to spend time outdoors with respect to warm climate. The semiopen shopping facility naturally creates leisure zones between shops. Open air patios offer a relaxing feeling under the shado.

Square & Park

广场和公园

抱树亭
Treehugger, One Fine Day

项目档案	Project Facts
设计：Holger Hoffmann	Design：Holger Hoffmann
项目地点：德国	Location：Koblenz, Germany
完成时间：2011	Year：2011

整个建筑由多边形框架构成，这种处理的目的是在建筑与周边环境之间创造一种微妙的联系。亭子毗邻 basillica st. castor，这里之前曾是一个停车场，亭子的结构与园艺展览和周边松散的城市肌理融为一体。

建筑外立面是透明的印刷图案玻璃，透明的轮廓模糊了室内外空间的边界，还将周边环绕的树木倒影叠加在建筑立面上。建筑材料选用的是胶合木板，园艺展结束后，这座建筑会被拆除并运往其它地方。

这个方案的平面采取了单向线性空间布局，一系列几何形的装置创造了一个带尖角的内部空间。设计师将三个独立的功能区合并到一起，试图将空间分配最大化，让各种活动能够同时举行。展厅区域的上部被延伸到天花板的树形柱子覆盖，同时还创造了若干小隔间作为学习空间，讲座和会议则在空间中心区域举行。

白天，这仅仅只是一个被树形结构装饰的建筑，可一到了晚上，整座建筑都发生了变化。建筑内部使用者的活动被环境逐一的虚拟，整座建筑随着室内光影和颜色的变化而变得丰富多彩。

The polygonal geometry and manifold symmetries of nearby St. Castor's stellar vault have been a major inspiration for the project. Together with a rotationally symmetrical order, a system of interdependent geometrical relations was defined that was resilient, yet rigorous enough to adapt to specific structural and functional needs. Furthermore the "branching" and inherent "porosity" of the trees' leafy canopy above has been abstracted into the similarly "porous" pentagonal and rhombic tessellation of the surfaces.

The pavilion uses the shape of an extruded pentagon instead of a square-box——a simple geometrical "plus" that changes the building's appearance significantly in relation to a beholder's standpoint. The partly screen printed glass curtain blurs the interior construction and superimposes it with reflections of the surrounding trees. The lighting changes drastically during the night. An interactive light-installation reacts to movements of the visitors, creating different visual effects.

广场和公园

Saint Georges 广场
Saint Georges Environment

项目档案

设计：Guerin & Pedroza 建筑事务所
项目地点：法国
面积：10 000 平方米

Project Facts

Design：Guerin & Pedroza Architectes
Location：Toulouse, France
Site Area: 10,000 m²

对设计师来说，这个项目的挑战性在于：一方面，在偌大的广场上同时设计东西方向和南北方向上的两条人行道，并为广场带来生机和活力。另外一方面，将购物拱廊设计得更加具有吸引力，并增加可达性，满足了场地以及周边的环境需要而变得更加迷人。

能够引起路人注意，并引导他们进入这个空间，这一点是非常重要的。一个大型的阶梯便恰到好处地起到了这个作用，引导路人来到这里。各种不同植被以及材料的选用增加了整个空间的节奏感。不同的区域不仅植物、材料不同，而且灯光和物种也不同。

这个空间有三个明显的入口，并且都在主要街道上。椭圆形的铝制锥体叠加在一起，并用玻璃覆盖，给人耳目一新的感觉。在空间的另外一面，拱廊的入口处上方有一个铜质的天篷，扩大了拱廊与周围环境的兼容性。

Square & Park

For the designers, the challenge consisted, on the one hand, in bringing this immense space crossed by two pedestrian ways (east-west and north-south) back to life and on the other, in making the shopping arcade more attractive.
Offering passers-by clear reference points is the principal watchword. A large stairway is an invitation to discover the place Occitane. There, a rhythm is set up by variations in the treatment of vegetables and minerals. The space is divided into several sectors: pergola, grove, esplanade, campo, recreation area, carpet, lane. Each is set off from the other by the choice of materials, lighting and species. There are three entries with direct access from the street open onto the Espace Saint-Georges. A volume composed of two elliptical aluminum cones and a glass roof beckons the city towards the main entry into the shopping centre. On the other side of the space, a door topped with a large copper awning offers a view across the arcade.

广场和公园

公园中的"拼凑图"
A Patchwork in the Park

设计：Cigler Marani Architects
项目地点：捷克布拉格
完成时间：2010 年

Design: Cigler Marani Architects
Location: Prague, Czech
Year: 2010

位于捷克布拉格公园的是一个巨大的办公综合体建筑，由 12 栋楼组成，此外还包括一个庭院，上班族们可以在长满绿草的山上，待在国际象棋板旁边或坐在混合座位区，享受休闲时光。

The Park is a large office complex in Prague, Czech Republic. The complex of 12 buildings is designed by Cigler Marani Architects, and includes a courtyard for office workers to enjoy some break time in a patchwork of grassy hills, chess boards, and a mixture of seating.

Square & Park

游乐场
Playground

Van Beuningenplein 游乐场
Van Beuningenplein Playground

项目档案

设计：Carve 景观设计公司
项目地点：阿姆斯特丹

Project Facts

Landscape Architecture：Carve
Location：Amsterdam

这个公共空间原本是一个停车场。阿姆斯特丹决定将停车场转到地下，将地上的空间转变成一个游乐场。原先的停车场因为汽车、栅栏以及糟糕的绿化，使得场地本身在周围的房屋等环境中消失了。而通过清除汽车和其他障碍物，这个空间才重新成为社区的一部分。同时为了方便居民，还在树篱旁设计了长椅和小型的花园。
一个彩色的充满生气的新地面将公共空间和私人空间之间的界限消退了不少。游乐场由四季长青的绿化带包围着，增添了游乐场与周围环境的协调性。游乐场的中心区域主要用于各种活动。在凹陷的运动场区域，还有戏水区以及冬天滑冰区。场地的整体表面呈波浪状，各种娱乐设施满足了各年龄段居民的需求。

Playground

游乐场

Parked cars dominate the public space both physically and functionally. Therefore the city of Amsterdam has decided to construct an underground parking garage at "Van Beuningenplein" at the existing playground. On top of the new parking garage, the former play and sport area had to return.
In the former situation, the "Van Beuningenplein" was hidden from view by cars, fencing and poorly maintained green. By eradicating the cars and other obstructions, the facades of the surrounding houses are connected to the square and the square becomes once again part of the neighbourhood. Along the facades, hedges are placed in strategic locations leaving space for resident initiatives, like a bench or a facades garden.

Playground

The boundary between private and public has become less rigid, and a colourful and lively plinth is the result. Green borders of perennials frame the central part of the square without isolating it from its surroundings. The central part is designated for sports and play. On the sport field, there is also space for a water feature and in wintertime ice-skating is possible. By placing special elements on the edge of the sunken sport field, a skatable edge has been created. The playing area is a large wavy surface with different playing towers, play elements for all ages and a water playground for summertime.

游乐场

 Zurichhorn 游乐场
The Zurichhorn Playground

项目档案

设计：Vetschpartner Landschaftsarchitekten AG
项目地点：瑞士，苏黎世
面积：4 700 平方米

Project Facts

Landscape Architecture：Vetschpartner Landschaftsarchitekten AG
Location：Zurich, Switzerland
Site Area：4,700 m²

整个游乐场由网状的栅栏构成，另一边是竹林。整个游乐场的地形类似一个沙滩，在空间上分为两层。上层是休息区和隔离区，下层是活动区域，主要有荡秋千和滑梯，其中最吸引人的是一个彩色的戏水区，戏水区旁边还有一个沙盒。在靠近街道的那边还有一堵曲线状的墙，这保持了游乐场的安静，另外这堵墙上还设计了各种用于攀爬的扶手。

The playground is framed by a fine meshed fence, which continues to Bellerivestreet into a bamboo grove. To the road side, a varied and curved wall, as noise abatement measure, ensures the quality of stay. It is equipped with various climbing. The surface of the entire play area appears similar to a beach as a sandy base. The square is divided into two levels. The upper level serves as the seat edge and divides the playing area. The lower level is a game framework with modular interchangeable dements, such as swing, hanging toy and slide. A colored water play is the attraction of the interactive animated game play side. Next to it is a large sandbox with different rocks arranged.

Playground

游乐场

火蜥蜴游乐场
Salamander Playground

项目档案

设计：Cardinal Hardy
项目地点：加拿大，蒙特利尔

Project Facts

Landscape Architecture: Cardinal Hardy
Location：Montreal, Canada

这个项目位于蒙特利尔皇家山公园。这个公园是由 Frederic Law Olmstead 在 1874 年设计的，每年的游客超过 300 万。这个项目会在原来公园规划的基础上加以改进，但是依然会保持公园的特色。方案中包括一个主题游乐场，一个拥有 30 个桌子的野炊区，重修的大道和小径，景观设计上还是以林地特色为基础。

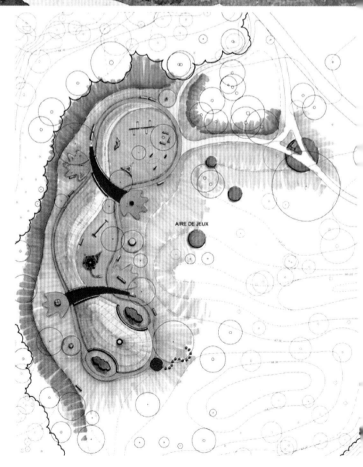

Playground

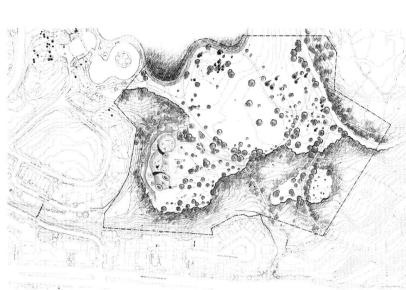

游乐场

游乐场的主题就是皇家山公园的两栖动物——蓝色斑点火蜥蜴。整个火蜥蜴形状区域包括了游乐设施、水景和其他设施，这些设施能够提高孩子们的主动性、认知能力和社交能力的发展。这个项目的目的是要在城市里建造一个有益于健康的大面积绿色空间，在垂直方向上突出与周围环境的和谐相融；在水平方向上更加衬托火蜥蜴的形状和鲜艳的颜色。整体结构上，色调主要采用中性和离散的颜色，显得若有若无。为了保持公园原有的自然生态，设计中尽量减少建筑面积。另外，在地面上采用自然的材料，可以使得表面的水渗透到土壤里。连接两个生态网络群落的生态走廊种上了当地的植物，从而充实了下层植物。

另外设计师还设计了一个"儿童权利"广场，用于对公众的教育。这个广场上有用于公众开会的海报栏，两本宣传册子和一张调查表，还有一个投影仪，并且对所有年龄阶段的人都适合，人们可以在这里畅所欲言，发表自己对儿童权利的看法。

With over three million visitors a year, Mont-Royal Park in Montreal was designed by Frederic Law Olmstead in 1874. Despite the evolution of the park, the essentials of the original plan remain. Given this heritage designation, a dozen municipal and provincial organizations had to ratify this project, which included: A play ground conceived with a theme derived from Mount Royal itself, a picnic area in a grassy plain with approximately thirty tables, the redevelopment of roadways and paths which reiterate Olmsteadian framed viewpoints as well as a renewed management of the landscape based on its woodland characteristics.

The theme is the Blue Spotted Salamander, an amphibian native to Mount Royal and the starring feature which organizes the play structures and other park elements. Water features and other innovative play structures are integrated into the silhouette of the salamander as it rises from the earth; this instigates a different kind of play, which encourages the children's motor, cognitive and social development. Beyond simply contending with a heritage site, the project highlights the therapeutic influence of this large scale green space in the city. The design was based on two distinct projections of the space; a vertically nuanced integration into the surrounding environment, and a horizontal plain contrasting the natural surroundings with the silhouette of the salamander and its bright colors. The neutral and discrete coloring of the structures allows them to melt when superimposed on the Olmsteadian decor. The built area is as limited as possible in order to keep the ecological footprint to a minimum and the use of natural materials on the ground let surface water percolate into the soil. At the edge of the clearing, an ecological corridor linking two main nodes of a local ecological network is planted with indigenous species in an effort to regenerate the understory.

Against this unusual backdrop, the landscape architect designed a Children Rights promenade of didactic elements. Public interpretive panels allow people of all ages to discover the rights guaranteed to children by the International Convention of Children's Rights. Coordinated by the landscape architect, three posters, two brochures, a questionnaire and a power point presentation are developed for public meetings.

游乐场

Playground

游乐场

Rommen 学校和文化中心
Rommen School and Cultural Center

项目档案

景观设计：tengen & Bergo AS
建筑设计：L2 Architects
项目地点：挪威，奥斯陆，Rommen
完成时间：2010

Project Facts

Landscape Architecture：tengen & Bergo AS
Architect：L2 Architects
Location：Rommen, Oslo, Norway
Year：2010

原先的学校教学楼太小，应该被取代。新的教学楼包括一个宽阔的多功能运动大厅，一个表演大厅和一个开放式图书馆。文化中心办公室和当地委员会也在这里。整个学校从一年级到十年级共有学生770名。室内和室外的设施将会为整个社区使用。

学校位于群山环绕的一个峡谷中，基地平整，北面稍微突起。学校的南面有一些山涧，长满草的斜坡和珍贵的植被将山涧点缀得甚是好看。两条输电线穿插其中，因为大的一条在学校的西面，所以，整个项目和景观设计主要集中在东面。东面的植物带是焦点，一直延伸到校园。

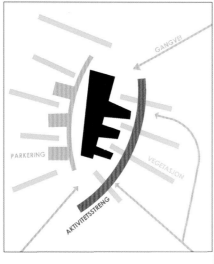

Playground

"广场"位于南面，在特殊的节日会吸引很多人。座位形状特别不同于以往的中规中矩，可以用于日常的户外会议。适合活动的区域主要集中在"活动带"，安静的活动则靠近教学楼。出入口，下车点和停车场位于北面和西面，下车点主要用于一至四年级的学生。

现有的通向学校的走道仍然保留，南面和东南面的通道被改进了。一条走道从南面延伸到整个方案的中心轴处。中心轴的东面是住宅区和活动区，西面是交通区和停车区。原先的通道很陡峭，因此不能满足日常的需求。新的通道从峡谷最低端的运动场通向学校，坡度是1:20。同时这条通道也可以被最小的学生使用。

学校教学楼两翼中间安静的地方有加高的木制甲板，这里可以被用作户外教室。孩子们可以坐在这里，甚至是躺着或在这里吃中饭。户外电脑数字化场地为老师和孩子们提供精彩的教学机会。场地上字母和人物的图像更无时无刻加强浓郁的教育氛围。在这个场地的外围有一个绿化区域，种有各种本土的和新移植的树木和灌木。树木上都有名字标识牌，以供孩子们了解挪威森林的物种。

建筑物的表面被景天属植物覆盖，从周围的房子和公路上依稀可见。景天属植物在全年会有不同的颜色。这块土地上的树木在有可能的情况下都被保留着。在学校的外部种上了大量的树木，草坪上春天的花儿开启了学校多姿多彩的时光。

户外的区域在设计时注重进入的便利性，残疾人都可以自由进入活动区域。休息区域增设了一个坡度为1:20的新走道。盲文导视线取代一般的导视线，方便盲人使用。

整个区域的绿化设计非常突出。一个巩固的绿色结构对于环境是非常重要的。通道和停车区域的地表水会流入水渠中，同时也灌溉了水渠中的植物。过多的水就会通过管道流入当地的小溪中。

The former school building was too small, and had to be replaced. The new building includes a large multipurpose sports hall, its own performance hall and an open library. Offices for the community culture school and part of the local council are also located here. The school houses 770 pupils from 1st to 10th grade. The facilities both indoor and outdoor will serve the whole community.

The school is located to an almost flat site, slightly rising to the north in a valley surrounded by hills. East of the site there are ravines with grassy slopes and valuable vegetation belts. Two power lines are crossing the area. With the largest line to the west, the project and the landscape design therefore pay more attention to the east. The vegetation belts in the east are reinforced and continue into the campus. Closer to the building, they get more cultured, and "finger-merged" with the building wings.

"The square" is to the south, gathers many people on special occasions and will be a nice and sunny meeting place with a variety of informal seating options. Zones for activities are located along the "activity belt" and zones for more quiet playing are closer to the building. There are a variety of seating options. Access, drop off and parking to the site are in the north and west. Drop off is mainly for transport of children from 1st to 10th grade.

Existing walkways to the school are maintained, and access from south and southeast is improved. A walkway from the south turns into the main axis of the plan. East of these axis are areas for residence and activities, and in the west are traffic areas and parking lots. Existing footpaths are steep, and do not satisfy the requirements of universal design. A new walkway with a gradient of 1:20 is therefore built from the sports grounds in the lowest parts of the valley and up to the school. This path can also be used by the youngest school children who do not want to pass the older kids on the way to their classroom.

The elevated wooden decks in the quiet zones between the wings of the building can be used as outdoor classrooms. The children can sit, lie down or eat lunch. The digital outdoor computer ground provides exciting opportunities for teaching and learning outdoors. Letters and characters are painted on the school grounds for outdoor learning. In the outer zones of the site is a green zone of existing trees and newly planted trees and shrubs. The trees are named with small signs so that children can learn the name and family of the common Norwegian forest species.

The large roof is highly visible from surrounding homes and roads. The roof is covered with vegetation and mats of sedum plants, and is a pleasant complement to the rolling surroundings. Sedum-mats have different colors throughout the year. Existing trees on the land have been preserved when possible. A lot of trees are planted in the outer part of the school. Spring flowers in the lawn provide a pleasant start to the grooving season.

Playground

The outdoor areas are adapted to the requirements of accessibility, and all activities will be available for the disabled. A new walkway with a gradient of 1:20 is put in through the recreation area. Tactile guide lines are established where natural leading lines cannot be used.
There has been a significant tree planting on the site, which reinforces the green structure in the area. The entire roof of the building is covered with sedum. A strengthened green structure has become positive for the environment. Surface water from the access and parking zones is lead into wet ditches with vegetation. At high rainfall ditches has an overflow drain. Water from the roof and drains is led in pipes to local streams.

游乐场

Gjerdrum 高中
Gjerdrum High School

项目档案

设计：Ostengen & Bergo AS, MNLA 景观设计事务所
项目地点：挪威，Gjerdrum
面积：56 000 平方米

Project Facts

Landscape architect: Ostengen & Bergo AS, landscape architects MNLA
Location: Gjerdrum, Norway
Site Area: 56,000 m²

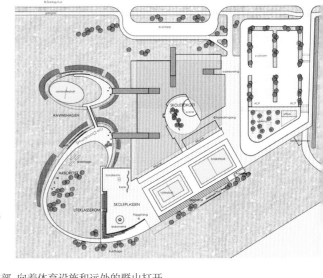

学校位于一个缓坡上，朝着西南侧，场地内有之前沟壑留下的印记。建筑的背部在北部，向着体育设施和远处的群山打开。学校的操场围绕建筑南侧的一个中心区域展开，西侧是一个形似峡谷的花园，它将建筑与木质的台阶相连，形成不同的活动空间。

建筑南面有一个活动区域，包括篮球场、排球场、跑道、沙坑和乒乓球台。这个区域同时也可以用作学生和老师聚集的地方。在其中一个沟壑里也有一个排球场。当然，这里也可以被用作户外教室。内部的庭院安静而荫凉。学校西南方的看台可以容纳很多学生。学校南面的墙体延伸成座位，学生可以在这里休息和上课。另外，这里也有一些较小的聚集区域，因此用于小组和大组学生的教学器材都有设计。南面的沟壑有一个草坪，学生们在这里尽情享受日光浴。小径一直通向植物园，那里有大量的挪威当地的树木。每一棵树上都刻有挪威语和拉丁语名字，还有高度和年龄。在沥青表面上漆有几何图形，方便学生们随时随地学习。

硬景观主要由沥青铺设，球场被混凝土路面包围着，平衡木游玩区的表面由黑色的橡胶沥青铺设。花岗岩路边石将铺设地面和走道分离开来。混凝土地面边缘采用钢筋作边框，加强了整个场地的线条美。草坪前的海棠花在秋天的时候为学校增色不少。树木按组栽植，无意间变成了汽车和单车的停车区。

Playground

The school is located in a gently sloping terrain towards the south west, with traces of past ravines. The building is located with its back to the north and opens against the sports facilities and the hills far away to the south. The schoolyard is developed around a central zone south of the building. To the west is built a garden formed as stylized ravines, linked to the building with wooden piers, including various activities. Along the south wall of the building is an activity area with basketball court, volleyball court, running track, jumping pits for long jump and table tennis. The area will also serve as a great gathering place for the school. In one of the ravines there is also a volleyball court. There are several possibilities for outdoor classrooms; the inner courtyard is a quiet, shielded room, and tribunes southwest of the school can take many students. The seating on the south wall of the building invites teachers and students to sit down and rest, but can also be used as an outdoor classroom. In addition there are smaller gathering places with a few steps, so that facilities for teaching in both large and small groups are present. The ravine in the south has a sunbathing lawn, and nature trail leading to the arboretum with all the Norwegian trees. Each tree is marked with Norwegian and Latin name, height and life expectancy. On the asphalt there are painted geometric shapes for learning.

游乐场

Playground

The hardscapes are mainly covered with asphalt. The courts are surrounded by a broad zone of concrete pavement. The ground cover under the balance play is black rubber asphalt. Kerbs of granite separate the pavement and roadway. Edges of steel frame zones with concrete pavement and tighten up the lines in the ravine garden. The ravine shape is shaped by precise slopes of lawn, in front of bush fields with Mains sargentii, which asssure spring blooming and autumn colors. Outside of the ravines the grass land are meadows. The trees are planted in groups. Car and bicycle parking has a lot of vegetation, both trees, shrubs and perennials.

达令港广场
Darling Quarter

项目档案

设计：ASPECT 工作室，Lend Lease, FJMT, David Eager, Hyder, Waterforms International
项目地点：澳大利亚，悉尼，达令港
完成时间：2011

Project Facts

Design：ASPECT Studio, Lend Lease, FJMT, David Eager, Hyder, Waterforms International
Location：Darling Harbour, Sydney, NSW, Australia
Year：2011

这个广场位于达令港东部，集贸易，零售和文化为一体。达令港是澳大利亚最受欢迎的一个游览胜地。这个项目最引人注目的一点就是它的世界一流的儿童游乐场，也是悉尼最大的游乐场。这个游乐场综合性很强，适合所有的儿童，为他们提供前所未有的冒险经历。订做的嬉水设施来自德国，在澳大利亚从来没有被使用过。通过新设计的行人街道，广场和城市的联系更加紧密，同时也增强了 Cockle 海湾以及其他区域的便民性。这个项目为原有的公共广场注入了新鲜的活力，提供了家庭聚会中心，游乐场和新的零售和贸易区，另外还有大面积的绿化区域。广场地面的材料选择成为公共空间设计的典范，主要选择高质量，简单和耐用的材料。

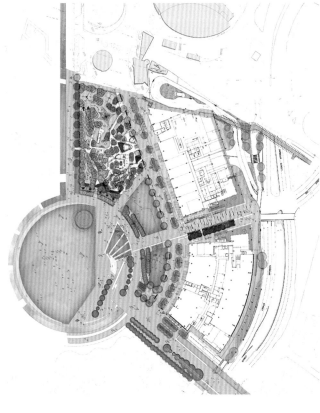

Playground

ASPECT Studio designed the public domain of Darling Quarter, a new commercial, retail and cultural development which sits at the eastern edge of Darling Harbour, the most visited tourist destination in Australia. The main attraction of this unique project is the world class children's playground, the largest in the Sydney CBD. The integrated play design creates an adventurous experience suitable for all abilities. The bespoke water play design elements are sourced from Germany, many of which have never been used in Australia. Through its new pedestrian streets, Darling Quarter builds a strong relationship with the city and reinforces the passage to Cockle Bay and beyond. The project revitalises the existing public open space. It provides a family focused core with an extensively redesigned playground, new retail and commercial precinct and generous grassed community greens which build on the successful public domain elements such as Tumbalong Park and the urban stream. Improved ground plane materials set a new benchmark for Darling Harbour, with an emphasis on high quality, simple, robust elements.

游乐场

PS1 跳舞的钢管
PS1 Pole Dance

项目档案	Project Facts
设计：SO-IL 工作室	Architect：SO-IL
项目地点：美国，纽约	Location：New York, USA
面积：1 400 平方米	Site Area：1,400 m²

SO-IL 设计公司的这个项目是 2010 年度青年设计师工程的获奖作品。这个项目旨在创造一个具有丰富感官体验的空间，而不是传统的固定形式。在探寻这个空间结构的时候，一方面要满足两个看起来似乎矛盾的功能要求——安静和喧闹，最重要的是要适合编舞的需求，而不是创造一个空间。在这个空间里，人类和结构有了新的概念关系。钢管和网相互交织在一起，人们活动和环境因素的变化都会改变它们的形状，得到新的平衡。这个网状结构具有很强的弹性，人们可以根据自身的需要自由创造活动，挑战自己的极限，或只是观赏跳舞。

主要空间的旁边有一个小庭院，在这里人们可以获得丰富的声音体验。其中有八条钢管安装有加速电子装置，这种装置会随着钢管位置的改变，将这种改变转换成音调。这样一来，人们在运动的时候，也可以欣赏到自己运动时所产生的独一无二的声音。

简单的材料和优雅的外观使之转变为一个游戏和社交的场所，虚拟世界中很多东西在这里实现。此外，这里还设有坐席、饮水处和酒吧区。所有的材料都采用简单的可持续材料，可以回收再生。

In this proposal, the designers take the opportunity to further contemporary explorations to create sensory-charged environments, rather than finite forms. Especially in the case of envisioning a temporary structure for the PS1 courtyard, which needs to perform two seemingly contradictory tasks (calming and carousing), a worthy proposition will need to consider the choreography of situations rather than object making. We designed a participatory environment that reframes the conceptual relation between humankind and structure. It consists of an interconnected system of poles and nets whose equilibrium is constantly affected by human action and environmental factors, such as rain and wind. Upon discovery of its elasticity, visitors engage with the structure to invent games, test its limits or just watch it gently dance.

Playground

The small courtyard adjacent to the main space holds an immersive and interactive portion where visitors can create and control a rich sound experience from the installation. Eight poles contain "accelerometers" electronic devices that measure the motion of the poles connected to custom software that converts motion into tones specifically composed for the installation. An iPhone application allows visitors to affect the quality of sound for each pole in real time. By turning the effects levels up or down, the audience can collaboratively vote to change the active sound of their environment.

The entire system is assembled from ready-made materials. The details allow for the system to be broken down without material degradation. Most components have been repurposed since the installation was closed.

游乐场

Norteland 游乐场
Norteland Playground

项目档案	Project Facts
设计：Studio Dass	Design: Studio Dass
项目地点：葡萄牙	Location：Portugal
完成时间：2011	Year：2011

本案是一个位于 Norte 购物中心室内的游乐场。设计灵感来源于彩虹、小山、房子、树木和湖泊。形象的游乐设施让孩子们可以尽情地玩耍，并鼓励孩子们探寻新的东西，对他们体力、认知和社交方面的发展都有很大的帮助。根据彩虹的颜色，设计师将整个游乐场分为七个主题区：紫色区域的主题是睿智；蓝色区域的主题是天空；青色的主题是水；绿色的主题是大自然；黄色的主题是灯光；红色的主题是速度。不同颜色的区域适合不同年龄阶段的孩子。紫色的区域适合年龄最小的孩子，而红色区域适合稍微大一点的孩子。

The project is an indoor playground for Norte shopping inspired by the rainbow, and the rounded shapes of Norte land resemble elements of landscape: small mountains, houses, trees, lakes. These abstract forms encourage kids to seek new meanings and images, also with the variety of games and activities, to stimulate the senses of children and foster the physical, cognitive and social development. The rainbow color palette divides the playground into seven themed areas, based on the meanings of each color: purple——wisdom; blue——sky; cyan——water; green——nature; yellow——light; orange——fire; red——speed. The games of each color suit different ages, and the violet color is for the youngsters and the red color for the elders.

Playground

游乐场

谷脊草甸游乐场
Valley Ridge Meadow Playground

项目档案 | Project Facts

设计：Museo Guggenheim Design: Museo Guggenheim
项目地点：西班牙，毕尔巴鄂 Location：Bilbao, Spain
完成时间：2010 Year：2010

本案集自治性、参与性、安全性和活动性为一体。在设计以及建造的过程中都有孩子们的参与，孩子们的需求和建议才是最好的设计原则与目的。这个游乐场是成年人运用自己的能力诠释儿童文化的代表。

The project supports autonomy, participation, safety, and mobility for children in public space. Through children's proposals and ideas, as well as participation in decision-making and project realization, the project builds an understanding of children's culture in the minds of adults.

Playground

图书在版编目（CIP）数据

公共景观集成.公共景观/广州市唐艺文化传播有限公司编著.-- 北京：中国林业出版社，2016.4

ISBN 978-7-5038-8442-9

Ⅰ.①公… Ⅱ.①广… Ⅲ.①景观设计-图集 Ⅳ.①TU986.2-64

中国版本图书馆CIP数据核字(2016)第050694号

公共景观集成　公共景观

编　　著	广州市唐艺文化传播有限公司
责任编辑	纪　亮　王思源
策划编辑	高雪梅
文字编辑	高雪梅
装帧设计	杨丽冰

出版发行	中国林业出版社
出版社地址	北京西城区德内大街刘海胡同7号，邮编：100009
出版社网址	http://lycb.forestry.gov.cn/
经　　销	全国新华书店
印　　刷	深圳市汇亿丰印刷科技有限公司
开　　本	220 mm × 300 mm
印　　张	19.75
版　　次	2016年8月第1版
印　　次	2016年8月第1次印刷
标准书号	ISBN 978-7-5038-8442-9
定　　价	316.00元（精）

图书如有印装质量问题，可随时向印刷厂调换（电话：0755-82413509）。